Scandalous Knowledge

Science and Cultural Theory

A Series Edited by Barbara Herrnstein Smith and E. Roy Weintraub

Scandalous Knowledge

Science, Truth and the Human

Barbara Herrnstein Smith

Duke University Press

Published in the US in 2006 by
Duke University Press
Box 90660, Durham, NC 27708-0660

First published in the UK in 2005 by
Edinburgh University Press Ltd
22 George Square, Edinburgh, Scotland

Typeset in Adobe Sabon by Servis Filmsetting Ltd, Manchester,
and printed and bound in Great Britain
by MPG Books Ltd, Bodmin, Cornwall

Library of Congress Cataloging-in-Publication Data

Smith, Barbara Herrnstein.
Scandalous knowledge : science, truth and the human / Barbara Herrnstein Smith.
p. cm. – (Science and cultural theory)
Includes bibliographical references and index.
ISBN 0-8223-3810-6 (cloth) – ISBN 0-8223-3848-3 (pbk.)
1. Constructive realism. 2. Knowledge, Theory of. 3. Science–Philosophy.
4. Postmodernism. I. Title. II. Series.
Q175.32.C66S65 2006
501–dc22

2005027127

Contents

For Stephen

Acknowledgements

Versions or portions of several chapters in the present book were published earlier, as follows: Chapter 3 is expanded from 'Netting Truth', *PMLA*, 115: 15 (October 2000), pp. 1,089–95; a slightly different version of Chapter 4, 'Cutting-Edge Equivocation: Conceptual Moves and Rhetorical Strategies in Contemporary Anti-Epistemology', was published in *The South Atlantic Quarterly* 101: 1 (2002), pp. 187–212, in a special issue, 'Vicissitudes of Theory', edited by Kenneth Surin; portions of Chapter 6 are drawn from 'Sewing Up the Mind: The Claims of Evolutionary Psychology', in *Alas, Poor Darwin: Arguments Against Evolutionary Psychology* (2000), edited by Hilary Rose and Steven Rose (London: Jonathan Cape Publishers), pp. 155–72; a slightly different version of Chapter 7, 'Animal Relatives, Difficult Relations', was published in *differences: A Journal of Feminist Cultural Studies* 15: 1 (2004), pp. 1–20, in a special issue, 'Man and Beast', edited by Ellen Rooney and Elizabeth Weed. Permission to publish these versions or portions is hereby gratefully acknowledged.

During the time this book was under way, a number of friends and old co-conspirators as well as colleagues and students at Duke University and, more recently, Brown University provided intellectual company and stimulation of a high order. Special thanks to Güven Güzeldere, Casper Jensen, Claudia Koonz, Javier Krauel, Hal Sedgwick, Julie Tetel and Roy Weintraub for valuable comments on individual chapters; to Colleen Lamos for significant invitations and mediations; to Tom Cohen, Susan Oyama, Andy Pickering and Arkady Plotnitsky for fortifying exchanges along the way; and to Stephen Barber for his careful review of the entire manuscript and for much else besides. The 1998 workshop on evolutionary psychology at Portrack, Scotland and the 2001–2 New Beginnings faculty seminar on science and technology studies at Duke provided welcome opportunities to test angles and sharpen formulations with specialists in various fields. A conversation in Copenhagen with Stig

Brorson and an exchange with Bruce Mazlish were important for the chapter on Ludwik Fleck. Students in successive versions of the course, '20th-Century Reconceptions of Knowledge and Science', and participants in the faculty seminars titled 'Scandalous Knowledge' at the Simon H. Rifkind Center for the Humanities, City College of New York, in 1998, and at the Humanities Institute, University of Illinois-Chicago, in 2001, required and exhibited high degrees of intellectual engagement and rigour in exploring these ideas and issues. Dominique Groenveldt fine-tuned my translation of some German texts. Preparation of the manuscript benefited greatly from the expert assistance of Zak Sitter and Sarah Lincoln.

Scandalous Knowledge

Chapter 1

Introduction: Scandals of Knowledge

It has been said that knowledge, or the problem of knowledge, is the scandal of philosophy. The scandal is philosophy's apparent inability to show how, when and why we can be sure that we know something or, indeed, that we know anything. Philosopher Michael Williams writes: 'Is it possible to obtain knowledge *at all*? This problem is pressing because there are powerful arguments, some very ancient, for the conclusion that it is not . . . Scepticism is the skeleton in Western rationalism's closet'.[1] While it is not clear that the scandal matters to anyone but philosophers, philosophers point out that it should matter to everyone, at least given a certain conception of knowledge. For, they explain, unless we can ground our claims to knowledge *as* such, which is to say, distinguish it from mere opinion, superstition, fantasy, wishful thinking, ideology, illusion or delusion, then the actions we take on the basis of presumed knowledge – boarding an airplane, swallowing a pill, finding someone guilty of a crime – will be irrational and unjustifiable.

That is all quite serious-sounding but so also are the rattlings of the skeleton: that is, the sceptic's contention that we cannot be sure that we know anything – at least not if we think of knowledge as something like having a correct mental representation of reality, and not if we think of reality as something like things-as-they-are-in-themselves, independent of our perceptions, ideas or descriptions. For, the sceptic will note, since reality, under that conception of it, is outside our ken (we cannot catch a glimpse of things-in-themselves around the corner of our own eyes; we cannot form an idea of reality that floats above the processes of our conceiving it), we have no way to compare our mental representations with things-as-they-are-in-themselves and therefore no way to determine whether they are correct or incorrect. Thus the sceptic may repeat (rattling loudly), you *cannot* be sure you 'know' something or anything at all – at least not, he may add (rattling softly before disappearing), if that is the way you conceive 'knowledge'.

There are a number of ways to handle this situation. The most common is to ignore it. Most people outside the academy – and, indeed, most of us inside it – are unaware of or unperturbed by the philosophical scandal of knowledge and go about our lives without too many epistemic anxieties. We hold our beliefs and presumptive knowledges more or less confidently, usually depending on how we acquired them (I saw it with my own eyes; I heard it on Fox News; a guy at the office told me) and how broadly and strenuously they seem to be shared or endorsed by various relevant people: experts and authorities, friends and family members, colleagues and associates. And we examine our convictions more or less closely, explain them more or less extensively, and defend them more or less vigorously, usually depending on what seems to be at stake for ourselves and/or other people and what resources are available for reassuring ourselves or making our beliefs credible to others (look, it's right here on the page; add up the figures yourself; I happen to be a heart specialist).

Another more formal way to respond to the scandal of knowledge is to set about refuting scepticism on logical grounds (most commonly as self-refuting or as entailing a manifestly unacceptable solipsism or relativism) and/or to observe that at least some forms of knowledge appear incontrovertible – for example, our appreciation of the truths of mathematics, or our consciousness of our own states of mind, or our experience of simple sensory facts (my hand is before my eyes, and so forth) – and hope or claim to go on from there. Making good on such hopes and claims has been the central project of Western epistemology from at least Descartes' time to the present. It has also been a preoccupation of philosophy of science since the beginning of the twentieth century.

A quite different but still formal way of responding is to accept the propriety of the sceptic's scepticism, given the classic conceptions of knowledge in play, but to radicalise it by extending that scepticism to those conceptions themselves: that is, to question, re-evaluate, and, as necessary, revise the system of ideas, definitions, distinctions, principles and undertakings in which the concept of knowledge or cognition has been classically situated in Western thought. This third way of handling the scandal of knowledge is at the centre of the present book, the specific topic of a number of its chapters and the perspective from which all of them are written. Accordingly, its title has a double referent: on the one hand, the more or less intractable theoretical problems presented by classic conceptualisations of knowledge; on the other, the development of significantly different – and, from many perspectives, scandalising – ways to think about knowledge and to investigate the phenomena and processes that appear to constitute it. (An additional referent will be noted at the end of this chapter.)

Classical epistemology is a definitively philosophical pursuit, normative (determining, directing, judging, justifying) in aim, logical-analytic in method. The alternative forms of knowledge-investigation in question here, notably constructivist epistemology and social studies of science in the tradition of Ludwik Fleck, Thomas S. Kuhn, Michel Foucault, David Bloor, and Bruno Latour,[2] have major theoretical components but are not conventionally philosophic enterprises. Their aims with regard to knowledge and science are largely descriptive and explanatory, not normative, and their methods, though certainly involving conceptual inquiry and (re)formulation, are largely empirical: archival and observational. In questioning classical conceptions of knowledge and pursuing the study of science as a set of historical phenomena and social-cultural practices, constructivist epistemology and post-Kuhnian science studies are radically different both from philosophical epistemology in the tradition of Descartes and Kant and from mainstream philosophy of science in the tradition of the Vienna Circle and Karl Popper. Moreover, the accounts of the emergence, stabilisation and transformation of scientific knowledge developed in these fields are radically different both from traditional lay ideas about scientific discovery, truth and progress and from the views of science held by many or most practising scientists. That is quite a lot, of course, from which to be radically different. It is not surprising, then, that the critiques, alternative approaches and alternative formulations developed in these relatively new fields have been misunderstood, misrepresented and greeted with ridicule, distress and/or outrage from so many quarters.

I

It will be useful to say a few words first about what I shall be referring to throughout the book as *constructivism*, including both constructivist epistemology and constructivist history and sociology of science. In most informed contemporary usage, including the usage of practitioners, the term 'constructivism' indicates a particular way of understanding the relation between what we call knowledge and what we experience as reality. In contrast to the understanding of that relation generally referred to as 'realism', constructivist accounts of cognition, truth, science and related matters conceive the specific features of what we experience, think of and talk about as 'the world' (objects, entity-boundaries, properties, categories and so forth) not as prior to and independent of our sensory, perceptual, motor, manipulative and conceptual-discursive activities but, rather, as emerging from or, as it is said, 'constructed by' those activities. In contrast to the prevailing assumptions of rationalist philosophy of mind, constructivist accounts of cognitive processes see *beliefs* not as discrete, correct-or-incorrect propositions about or mental representations of the world but, rather, as linked perceptual dispositions and behavioural routines that are continuously strengthened, weakened and reconfigured through our ongoing interactions with our environments. In contrast to referentialist views of language, constructivist accounts of *truth* conceive it not as a matter of a match between, on the one hand, statements or beliefs and, on the other, the autonomously determinate features of an altogether external world (Nature or Reality), but, rather, as a situation of relatively stable and effective mutual coordination among statements, beliefs, experiences and practical activities. And, in contrast to logical positivist or logical empiricist views, constructivist accounts of specifically *scientific* truth and knowledge see them not as the duly epistemically privileged products of intrinsically orthotropic methods of reasoning or investigation ('logic' or 'scientific method') but, rather, as the more or less stable products of an especially tight mutual shaping of perceptual, conceptual and behavioural (manipulative, discursive, inscriptional and other) practices in conjunction with material/technological problems or projects that have especially wide cultural, economic and/or political importance.[3]

Constructivism is often conflated with 'social constructionism'. Distinguishing between the two is difficult because both terms have shifting contemporary usages and variants ('construct*ionism*', 'social construct*ivism*', and so forth) and because the views and enterprises they name have complex intellectual-historical connections. Nevertheless, their simple identification obscures significant differences of origin, emphasis and intellectual or ideological operation. As a set of ideas concerning the nature of knowledge and operations of human cognition, *constructivism* has been developed largely by epistemologists, psychologists, cognitive scientists, and historians, philosophers and sociologists of science. Key proto-constructivists include Nietzsche and Heidegger; influential mid-twentieth-century figures include Peter Berger and Thomas Luckmann, Jean Piaget, Paul Feyerabend and Nelson Goodman; important contemporary constructivists include Susan Oyama in developmental psychology and Gerald Edelman, Francisco Varela and Antonio Damasio in cognitive science.[4] All these theorists are or were interested in the processes and dynamics of cognition: either microcognition, that is, individual learning, knowledge and perception, and/or macrocognition, that is, intellectual history and the cultural-institutional-technological operations of science. Accordingly, their arguments are or were primarily with other scholars and theorists of cognition and science, especially though not exclusively philosophers.

With regard to ideas of knowledge and science, *social constructionism* is a more culturally focused and politically engaged – or, as it is variously claimed or complained, 'critical' – set of views. As such, and especially as maintained by cultural critics, feminists, gender theorists and other scholars and critics in connection with such problematic practices as racial classifications, gender bias or normative heterosexuality, it operates primarily as an effort to challenge relevant beliefs – including those offered by scientists or in the name of science – by denaturalising them, revealing their dependence on historically or culturally particular discursive practices and/or exposing their implication in the preservation of prevailing social and political arrangements. In cultural-studies usage, the phrase 'socially constructed' commonly contrasts with such status-quo-justifying notions as 'natural', 'innate', 'proper' or 'inevitable'. In science-studies usage, it commonly contrasts with such realist-rationalist notions as 'pre-existing in Nature or Reality' and/or 'accepted for good reasons'. The idea of the *social* figures centrally for both constructivist theorists and social-constructionist critics, but in diverse (though not necessarily mutually exclusive) ways. Thus a constructivist sociologist of science is likely to stress the 'social' – here in the sense of collective, intersubjective and/or institutional – aspects of scientific knowledge as opposed to individualistic conceptions of the knower or the scientist, while the social-constructionist gender theorist is likely to stress the 'social' – here in the sense of class-related, culturally mediated, economic and/or political – forces involved in gender distinctions in opposition to socially conservative defences or biologistic accounts of those distinctions.

The citation of Foucault occupies a hinge position with regard to these confusions or conflations (although, as it happens, neither term – 'constructivism' or 'socially constructed' – appears in his writings, at least not centrally). A number of Foucault's earlier works, especially *Madness and Civilization*, *The Order of Things* and *The Archaeology of Knowledge*, develop models of the dynamics of intellectual and institutional history, including various human sciences and academic-professional disciplines, that are quite congruent with those of the constructivist historians and sociologists mentioned above. For example, Foucault's 'epistemes' (and, later, 'discourses') are comparable along many lines to Fleck's 'thought styles' and Kuhn's 'paradigms'. Although the emphases are different in each case, all three point to the existence of conceptual-discursive systems that both enable and constrain the processes of cognition – perception, classification and so forth – for the members of some historically or otherwise specific collective (for example, nineteenth-century clinical psychiatrists in Foucault's account, early twentieth-century medical pathologists in Fleck's, or seventeenth-century chemists in

Kuhn's). At the same time, a number of Foucault's later works, especially *The History of Sexuality* and the essays and interviews collected as *Power/Knowledge*, have been key reference points for the notion of social construction, both for those who appeal to it in challenging conventional assumptions about some human trait, category or institution and for those who dismiss the notion as an absurd ideology-driven denial of the self-evident given-ness (for example, innateness or universality) of the trait, category or institution in question. Thus, as we shall see in Chapters 6 and 7, casual dismissals of Foucault along with what is represented (often quite ignorantly) as 'social constructionism' figure recurrently in the self-promotions of evolutionary psychology.

II

In *The Social Construction of* What?, Ian Hacking observes that nominalism is a crucial conceptual commitment of constructivist epistemology (which, as it happens, he calls 'social constructionism').[5] Hacking explains nominalism as the 'den[ial]', contra realism's affirmation, that Nature is inherently structured in certain ways.[6] Contrary to his implication, however, constructivists do not characteristically 'deny' *metaphysically* what realists evidently metaphysically maintain: namely, first, that Nature *is* structured in certain ways inherently (meaning independent of our perceptions, conceptions and descriptions) and, second, that we properly assume (Hacking says 'hope') that those ways are largely in accord with our perceptions, conceptions and descriptions of them. Rather, constructivists typically decline, in their historical, sociological or psychological accounts of science and cognition, to presume either any particular way the world inherently is *or* such an accord. This professional ontological agnosticism is not, as realists may see it, a perverse refusal of common sense but an effort at due methodological modesty and theoretical economy.

Nevertheless, Hacking's observation of an association between constructivism, thus labelled, with nominalism, more appropriately understood, is correct and suggests other important intellectual connections and divergences. Specifically, nominalism (traditionally, the view that 'universals' are only names given to 'particulars') can be seen as an active, ongoing acknowledgement of the theoretical significance of the historicity of language. It is, in that sense, allied with Nietzschean and Foucauldian genealogy and with Derridean deconstruction, all of which attempt, among other things, to indicate historical and other ranges and shifts in the actual usages and understandings of theoretically fundamental terms: *truth*, *reason*, *power*, *good*, *nature*, *culture* and so forth. In constructivist accounts of knowledge and science, nominalism manifests itself as the ongoing questioning of standard understandings and treatments of such terms as *fact*, *discovery*, *evidence*, *proof*, *objectivity* and, of course, *knowledge* and *science* themselves. To question – that is, to decline for take for granted – conventional theorisations of such terms and concepts is not, of course, to attack, abandon or deny the existence of that which is called 'fact', 'discovery', 'evidence', 'proof' or 'objectivity'. One says 'of course', but that disidentification is by no means always obvious. To those who regard theoretical terms as denoting discrete entities or classes with specific, determinate, necessary-and/or-sufficient properties, the idea of a *reconceptualisation* of fact or knowledge will be hard to grasp. To them it would appear that there can only be proper or improper uses of such terms, only adequate or inadequate understandings of such concepts. Indeed, attempts to determine just such proprieties and adequacies with regard to just such terms and concepts (what is genuine knowledge? what are the criteria for genuine scientificity? what conditions are necessary and/or sufficient to warrant a claim of proof, validity or objectivity? and so forth) are among the central normative activities and missions of classical epistemology and of logical positivist/empiricist (and, later, analytic/ rationalist) philosophy of science. Clearly, the framing and pursuit of such missions involve assumptions about how terms, concepts and language generally operate and, thereby, about the more general dynamics of human cognition, verbal behaviour and intellectual history. All these matters, however, are centrally at issue in contemporary debates about science and knowledge. Accordingly, such attempts at definitive definition or demarcation – whether by stipulation, as in Popper's proposal of empirical falsifiability as a maximally suitable criterion of scientificity, or by 'conceptual analysis', as in contemporary analytic epistemology[7] – can be seen either as question begging or as otherwise (empirically, conceptually and/or pragmatically) problematic.

A crucial component of constructivism, linked to its nominalism, is the idea that terms and concepts operate as elements of larger systems or *networks* (the term is recurrent) of assumptions, beliefs and conceptual-discursive practices that are both densely internally interconnected and, for that reason and others, powerfully normative. Thus such technical terms as *planet*, *element*, *organ*, *disease*, *race*, *gene* or *intelligence* – and, similarly, *knowledge*, *science*, *reason* or *reality* – are seen by Fleck, Kuhn, Feyerabend and in post-Kuhnian science studies generally as having force and meaning not in individual, discrete and fixed relation to particular referents but as parts of historically and culturally specific systems of beliefs and practices. Kuhn sees the formation of such disciplinary conceptual-discursive-pragmatic systems as requisite for the pursuit of what he calls 'normal' science and their radical breakdown and wholesale replacement as the major events of intellectual history. Such systems have also been called – by, among others, Heidegger, Wittgenstein, Quine, Derrida and Lyotard – 'world pictures', 'language games', 'webs of belief', 'closures' and 'regimes of truth'.

There are a number of key points here. One is the interrelatedness and high degree of mutual determination of *conceptual-discursive* elements (ideas, definitions, distinctions, predications and so forth) and both *perceptual-cognitive dispositions* (observations, classifications, interpretations and so forth) and *material practices* (measurements, manipulations, the design and manufacture of instruments and so forth). Another is the *social* constitution and maintenance of such systems: that is, the fact that they arise from and are stabilised by ongoing verbal, cognitive, social and pragmatic interactions among the members of particular communities. A third point, accordingly, is the *contingency* of the epistemic viability of the systems and their elements: notably, the dependence of the specific meaning and force of individual terms and concepts ('disease', 'planet', 'quark', 'gene' and so forth) on their intersystemic linkages to other ideas, assumptions and related practices and, correspondingly, the dependence of what would be experienced as the adequacy or propriety of any of these ideas, assumptions and practices on the currency of the relevant system in the relevant community.[8]

In debates over such ideas since the 1960s and 1970s, the constructivist views outlined above have been interpreted – and, accordingly, assailed – as implying an everything-is-equally-valid relativism, 'anything goes' in the practice of science, and irrationalism in the choice and adoption of scientific theories. What those views certainly do imply is the conceptual and empirical inadequacy of prevailing philosophical accounts of scientific method, truth and progress.[9] Indeed, one way to understand the self-scandalising perceptions and interpretations of constructivism by mainstream philosophers is as a testament to the continuing coherence and normative power of the classic conceptual system, at least within the orbits of their particular epistemic communities.[10]

III

The specifically disciplinary distinctions evoked here – that is, between, on the one hand, academic philosophy and, on the other, the various fields (sociology, history, psychology and so forth) that study science empirically and/or offer naturalistic accounts of human cognition – play out in quite complex ways over the course of the twentieth century. Of particular interest in the later chapters of this book will be the operation of such distinctions in the 'science wars' of the 1990s, in the more general ideology of 'the two cultures', and in the claims of evolutionary psychology (contra other social-science disciplines) to exclusive scientific warrant in the explanation of human behaviour and culture. Of related interest, examined in the final chapter, are the different – sometimes conflicting, sometimes intersecting – ways that naturalistic and humanistic disciplines and discourses (ethology, primatology, formal ecological ethics, the animal rights movement, sociobiology and so forth) handle the increasingly complex question of the relation(s), theoretical and ethical, between humans and animals.

Some intellectual-historical background to these conflicts may be noted. Throughout the twentieth century, rationalist philosophy, while granting some measure of recognition to the empirical study of knowledge in the history and sociology of science and to experimental psychology in the study of cognition, has preserved the honour and centrality of its own epistemological enterprise through a series of anti-heretical campaigns and strenuously maintained distinctions. The major campaign here has been against 'psychologism', that is, the error of thinking that mathematics, mind, consciousness, language or any comparable set of phenomena or problems could be illuminated by laboratory studies of human behaviour or physiology – or, as it is said, 'reduced to' the findings of such studies. Demonstrating the gravity of that error was an important project at the turn of the century, pursued most tenaciously and successfully, at least in the eyes of fellow logicians and philosophers, by Frege and Husserl, thereby colouring both the analytic tradition of philosophy of mind (following Bertrand Russell and early Wittgenstein) and the phenomenological tradition (following Heidegger).[11] Although the campaign against psychologism is no longer pursued as such, its effects and latter-day counterparts are evident in ongoing debates over the possible contributions (if any) of neuroscience to the understanding of thought, cognition, feeling, reasoning or consciousness[12] and, as examined in Chapter 7, the complex logical-ideological stances of various humanist-rationalist philosophers in current debates over the animal-human connection (or divide) and the naturalising discourses of sociobiology and evolutionary psychology.

Of a piece with anti-psychologism is the compound distinction, drawn initially in the 1930s, between the 'context of discovery' and the 'context of justification' of scientific facts or theories, which distributes matters neatly if dubiously within science and neatly if invidiously between philosophy and other enterprises.[13] As the distinction is commonly explained, whereas the context of *discovery* of a scientific theory may involve such matters as personal interests or social influences and be of interest accordingly to biographers, historians or sociologists, its context of *justification*, which consists of the determination of the scientific legitimacy, conceptual coherence and objective validity of the theory in question, involves only matters of reasoning and is thereby properly the task of logicians and philosophers. A cognate distinction came to be drawn between 'internalist' and 'externalist' histories of science. The idea here is that *internalist* histories, which document the unfolding of scientific knowledge from the evidentiary-based solution of strictly scientific problems arising in and from the pursuit of science itself, may, as such, be useful to philosophy, but that *externalist* histories, which concern themselves with various 'outside' influences or factors (institutional or political conditions, technological projects and so forth) that occasionally deflect science from its essential epistemic aims (or so it is said) are, accordingly, without interest for philosophy.

Questions may be and have been raised regarding the propriety of all these distinctions. Is there a clear temporal and modal difference, for example, between scientific discovery and justification or, in effect, between scientists' accounts of phenomena and (their own) assessments of the adequacy of those accounts? Are philosophers especially well equipped to make assessments of scientific legitimacy or validity? Do scientific events ever unfold purely internally to science, and are external forces intrinsically alien to the furthering of scientific goals? Are we sure, for that matter, that personal interests, institutional conditions and technological projects are 'outside' science, or might they be essential aspects of science and, indeed, necessary for its functioning in the ways commonly valued? Are psychologists, historians and sociologists necessarily to be seen as upstarts and interlopers in examining the ways scientific theories are formed, assessed, stabilised, justified and transformed, and are their findings intrinsically without interest for philosophers? Or, rather, as all these questions suggest, may it not be the case that the strict distinctions and divisions of labour that underwrite philosophy's self-honouring role in the study of science and knowledge beg all the relevant questions and, where maintained, insure philosophy's self-confinement? That, in any case, became a key issue among philosophers themselves[14] as well as between philosophers and scholars or scientists in other fields in the second half of the century.

IV

As distinct from rationalist conceptualisations, knowledge (everyday, expert or scientific) may be understood not in opposition to ('mere') belief but *as* beliefs that have become relatively well established. So understood, knowledge/beliefs can be seen as emerging continuously from three interacting sets of forces: individual perceptual-behavioural activities and experiences; general cognitive processes; and particular social-collective systems of thought and practice. Accordingly, the familiar contrast between 'duly compelled by reason' and 'improperly influenced by interests and/or emotions' gives way to the idea that all beliefs are contingently shaped and multiply constrained. Similarly, such familiar distinctions as those between 'objectively valid scientific knowledge' and 'personal opinion' or 'popular superstition' are replaced by the idea that all beliefs are more or less congruent with and connectible to other relatively stable and well established beliefs; more or less effective with regard to solving current problems and/or furthering ongoing projects; and more or less appropriable by other people and extendable to other domains of application. That is, the differences expressed by the classic contrasts – and there certainly are differences, and they certainly are important – are not denied or flattened out (reason is not 'abandoned' for 'irrationalism'; scientific knowledge is not 'equated with' myth or ideology; and so forth), but are reconceived as variable gradients rather than fixed, distinct and polar opposites.

The chart below (see p. 12) summarises a number of these points and indicates some distinctive concepts and major foci of interest in contemporary research and theory regarding knowledge and science. I would stress that the two columns here are not to be taken as contrastive oppositions or as simple substitutions of one by the other but as differences or alternatives of various kinds. In a number of cases (for example, *communal, social, institutional* versus *individual*, or *activities, skills, practices* versus *propositions, laws, models*), the contemporary alternative is an addition to or expansion of the classic attributes and/or a shift of focus in the objects of research and theorisation. In other cases (for example, *embodied* versus *mental*, or *coordination* versus *correspondence*), the alternatives represent reconceptualisations or redescriptions of the classic notions but not their simple rejection.

A glance at this chart makes obvious enough the reasons for the recurrent misunderstandings, misrepresentations and sense of scandal we are remarking here. For in virtually every instance, the constructivist-pragmatist-interactionist alternative, no matter how theoretically well elaborated, empirically well documented, or conceptually coherent and

20th-Century Reconceptions of Knowledge and Science

Classic Realist, Rationalist, Logical Positivist Concepts	*Distinctive Constructivist, Pragmatist, Interactionist Concepts*
Individual	Communal, social, institutional
Interior, intellectual, mental	Exhibited, embodied, enacted
Propositions, laws, models	Activities, skills, practices
Representation, correspondence	Interaction, coordination
Discovery	Construction
Reason, logic, experiment	Negotiation, rhetoric, performance
Unity, progress	Multiplicity, transformation
Truth	Effectivity
Autonomy	Connection, interdependence
Objectivity	Interests
Transhistorical, universal	Historical, situated

practically effective from the perspective of those proposing it, will appear to those operating effectively with traditional conceptions either as a perverse reversal of a fundamental attribute of genuine knowledge or as a cynical dismissal of a fundamental value or ideal of science. As is generally the case with such head-on collisions of thought styles, neither decisive rebuttals nor rational resolutions, as these are commonly understood, are to be looked for. And, indeed, most debates on these issues, where they have intellectual content at all, consist largely, on the revisionist side, of functionally inaudible (as if never uttered) or invisible (as if the pages were blank) repetitions of central arguments and, on the traditionalist side, of the rehearsal of self-affirming defences and refutations, while, as discussed in Chapter 4, proposed 'middle ways' tend to be rightward-leaning, fundamentally unstable assemblies of contradictory views.

The situation described here is not, in my view, the herald of a major 'paradigm shift' in how we understand knowledge and science. The work of such twentieth-century historians of science as Alexandre Koyré, Kuhn and Foucault has made it difficult, of course, to conceive the dynamics of the history of science as cumulative progress toward, or unfolding of, manifest epistemic destiny. Series of radically discontinuous junctures, however, or recurrent cycles of plateaus, crises and revolutions, are not the only alternatives. Paradoxically, both are too static and monolithic when it comes to representing intellectual history in a general sense – that is, as distinct from the history of the natural sciences as such. A finer-grained and more dynamic model is suggested by Fleck's image of a 'network in continuous fluctuation' as adumbrated by the notions of 'strata', or multiple sites of activity and change, and 'polytemporality', or variable time scales, elaborated by Foucault and Latour.[15] In accord with such a model, one may conceive the dynamics of intellectual history, including the history of philosophy and of the other academic disciplines as such, as a continuous play of elements, sometimes contending, sometimes convergent, sometimes parallel, in which new elements (clusters of intertwined ideas, constructs, discourses and institutional practices) 'merge', in Fleck's words, 'with old ones to create stable points, which, in turn, are starting points for new lines everywhere developing and again joining up with others',[16] and so forth – without end, either as completion or determined destination, but not without event, development or differentiation.

If we see intellectual history this way, twentieth-century epistemology (broadly understood here) can be regarded as a period of especially high volatility – challenges, innovations, resistances, reformulations – in a field of especially numerous, mobile, energetic elements and complexly connected lines and nodes. What *can* be looked for, then, and seem already to be in view, are transformations, more or less radical and more or less widespread, among all the theoretical discourses, styles and enterprises involved, those of epistemology proper, those of rationalist-analytic philosophy of science, those of constructivist theory, and those of such related fields as cognitive theory, cultural theory and political theory: multiple ongoing challenges and responsive innovations, not across-the-board switches; mutual modification, appropriation and intermingling, not unilateral triumph.[17] Continued resistance along with misunderstanding, misrepresentation and self-scandalised reaction are as much to be expected as anything else. The difference between intellectual history and the history of science is that, in the former, where there are no technological-pragmatic pressures for closure and thus no Latourian 'black boxes' (that is, highly stabilised and taken for granted ideas or machines), it remains possible for mutually incompatible conceptual systems and thought styles (for example, rationalism and pragmatism) to coexist and contend with each other virtually ad infinitum, with the only victories – and highly unstable ones at that – being institutional. But, of course, such victories are not insignificant to those involved.

V

If scepticism is the skeleton in rationalism's closet, relativism is the bugbear under its bed. The topic or phantom haunts twentieth-century epistemology proper and figures centrally in the reception of constructivist views of knowledge and science. Relativism is approached in the chapters that follow from a number of angles. Chapter 2 attempts both to dispel a spectral 'postmodern relativism', noting its current potent invocation but thoroughly straw construction in intellectually conservative quarters, and to draw attention to a substantial *pre*-'postmodern' relativism: specifically, to the emergence, early in the century, of important critiques of and alternatives to traditional absolutist, objectivist, universalist accounts in a number of fields. The familiar association of relativist views with the idea that all theories (judgements, opinions and so on) are equally valid is examined critically in Chapter 3 in connection with 'network' accounts of the stabilisation of scientific knowledge. Chapter 4 is concerned with efforts by feminist epistemologists and other contemporary theorists to navigate conceptually and rhetorically between, on the one side, their attraction to various constructivist views and, on the other, their anxieties about the supposed intellectual, moral or political perils of an egalitarian relativism allegedly entailed by those views.

A few points may be noted here regarding that alleged entailment, which is invoked recurrently in discussions of the political and ethical implications of constructivist epistemology and science studies or, as it is typically put in these connections, of the 'denial' of commonsense (or rationalist-realist) accounts of truth and knowledge. Contrary to the recurrent identification of constructivism with an egalitarian (everything-is-equally-valid) relativism and, thereby, with a do-nothing, judge-nothing quietism, constructivist understandings of truth and knowledge, as outlined above, would not render one unable or unwilling to compare or judge divergent truth- or knowledge-claims: for example, those of Western biomedicine versus those of indigenous healers, or Darwinian evolutionary theory versus the idea of Intelligent Design. To be sure, one would not, given such understandings, be inclined to proclaim the absolute, objective or universal validity of any of these (or any other) claims. But one would certainly be inclined and equipped to affirm the superiority of one such claim over the other with regard to the various epistemic dimensions indicated above (congruence, connectibility, effectiveness, appropriability, extendibility and so forth) and to argue the relevance of those dimensions to the purposes at hand (handling a new epidemic disease, challenging the policy of a school board in court) and/or to more general ends and interests. The idea that such judgements and justifications, in rejecting or foregoing absolutist, objectivist or universalist appeals, would be rationally indefensible and/or rhetorically ineffective is a rationalist prejudice and a factual error.[18] What sustains the prejudice and error is the normative force of the classic conceptual system itself, which is to say, the continuing ideological and institutional dominance, in many quarters, of some form of rationalist-realist-positivist epistemology. As also indicated above, however, nothing in a contemporary constructivist understanding of intellectual history would lead one to claim or expect the simple overthrow of that epistemology.

Finally, note is taken in the pages that follow of the occurrence of 'relativist' – along with 'postmodernist', 'anarchist', 'Parisian', and other such terms – as a random epithet of derogation and dismissal in the 'science wars', culture wars, theory wars and discipline wars, particularly as invoked by traditionalist philosophers and scientists dealing with the intellectual developments traced here and by evolutionary psychologists seeking to discredit critics or rivals. Many of the philosophers, scientists and other academics producing such epithets in these connections are, in their respective fields, quite eminent. What is notable in virtually all the cases cited here, however, is their exceedingly limited first-hand acquaintance with, and, in fact, considerable ignorance of, the ideas thus derogated and dismissed – an ignorance especially remarkable where the issues are posed as matters of, precisely, intellectual competence, rigour and responsibility. Indeed, it's rather a scandal.

Notes

1. M. Williams, *Problems of Knowledge: A Critical Introduction to Epistemology*, pp. 2, 5.
2. Key texts here are Fleck, *Genesis and Development of a Scientific Fact*; Kuhn, *The Structure of Scientific Revolutions*; Foucault, *The Archaeology of Knowledge*; Bloor, *Knowledge and Social Imagery*; Latour, *Science in Action*.
3. For examples and elaborations of the various constructivist views described here, see Knorr-Cetina, *The Manufacture of Knowledge: An Essay on the Constructivist and Contextual Nature of Science*; Pickering, *Constructing Quarks: A Sociological History of Particle Physics*; von Glasersfeld, *Radical Constructivism: A Way of Knowing and Learning*; B. H. Smith, *Belief and Resistance: Dynamics of Contemporary Intellectual Controversy*, pp. 23–51; Christensen and Hooker, 'An Interactionist-Constructivist Approach to Intelligence'; and Chapter 3, below. For an instructive historical account, see Golinski, *Making Natural Knowledge: Constructivism and the History of Science*.
4. See, for example, Berger and Luckmann, *The Social Construction of Reality: A Treatise in the Sociology of Knowledge*; Piaget, *Biology and Knowledge: An Essay on the Relations between Organic Regulations and Cognitive Processes*; Feyerabend, *Against Method: Outline of an Anarchistic Theory of Knowledge*; Goodman, *Ways of Worldmaking*; Oyama, *The Ontogeny of Information: Developmental Systems and Evolution*; Edelman, *The Remembered Present: A Biological Theory of Consciousness*; Varela, Thompson and Rosch, *The Embodied Mind:*

Cognitive Science and Human Experience; Damasio, *Descartes' Error: Emotion, Reason, and the Human Brain*.

5. Hacking, *The Social Construction of* What?, pp. 82–4. Offering to clarify confusions, Hacking often extends them here. Although his historical placements and framing of key issues in the book are careful and instructive, Hacking's acknowledged ambivalence on those issues and, it appears, fundamental commitment to realist-rationalist understandings (or as he sometimes calls it, 'philosophical purism' [see, for example, p. 81]) lead him to oversimplify a range of constructivist ideas. For a comparable recent effort, historically informative but ambivalent (or inconsistent) in attitude and assessment, see Zammito, *A Nice Derangement of Epistemes: Post-Positivism in the Study of Science from Quine to Latour*. For a good overview and analysis of current positions on the constructivist-realist divide among philosophers of science, see Rouse, 'Vampires: Social Constructivism, Realism and Other Philosophical Undead'.
6. Hacking, *Social Construction*, p. 82.
7. See, for example, M. Williams, *Problems of Knowledge*, cited at the beginning.
8. For detailed historical examples, see Fleck, *Genesis*; Latour, *The Pasteurization of France*; Pickering, *The Mangle of Practice: Time, Agency, and Science*.
9. Empirical inadequacy would remain a key point. The views of Fleck, Kuhn and Feyerabend were shaped centrally by their own archival studies in the history of science and informed by current experimental studies of visual perception. Fleck's were also informed and illustrated by his observations as a practising scientist. As discussed below, however, rationalist philosophers of science dismissed appeals to historical data or empirical studies as irrelevant to the crucial *philosophical* questions regarding science.
10. Isabelle Stengers observes that Kuhn's account of paradigms, revolutions and normal science disturbed the conception of science among many philosophers of science but not among most scientists themselves. This, she notes, is because the essential thing for the latter, namely, the autonomy of the norms of the scientific community, was preserved (Stengers, *The Invention of Modern Science*, pp. 4–6). On the strained encounters between rationalist philosophy of science and constructivist history and sociology of science, see also B. H. Smith, *Belief and Resistance*, pp. 125–52.
11. See Notturno (ed.), *Perspectives on Psychologism*; Kusch, *Psychologism: A Case Study in the Sociology of Philosophical Knowledge*; Jay, 'Modernism and the Specter of Psychologism'.
12. See, for example, Bennett and Hacker, *Philosophical Foundations of Neuroscience*, and the debates over the 'eliminative materialist' views of Patricia Churchland and Paul Churchland in McCauley (ed.), *The Churchlands and Their Critics*. The phenomenological tradition has split into a number of diverse strands, including both strong self-differentiation from cognitive science (see, for example, Ricoeur's side of the dialogue in Changeux and Ricoeur, *What Makes Us Think?*) and significant alliances with it (see, for example, Petitot et al. (eds), *Naturalizing Phenomenology: Issues in Contemporary Phenomenology and Cognitive Science*).

13. The distinction, drawn in those terms by Hans Reichenbach, was given force in Popper's *The Logic of Scientific Discovery*. For a good brief discussion, see Nickles, 'Discovery'.
14. See, for example, Quine, 'Epistemology Naturalized'; Goodman, *Ways of Worldmaking*; Rorty, *Philosophy and the Mirror of Nature*.
15. Fleck, *Genesis*, p. 79; Foucault, *Archaeology*; Latour, *Pandora's Hope: Essays on the Reality of Science Studies*.
16. Fleck, *Genesis*, p. 79, translation slightly modified. The passage is discussed further in Chapter 3.
17. For recent examples of such transformation, appropriation and intermingling in the fields mentioned, see Longino, *The Fate of Knowledge*; Godfrey-Smith, *Theory and Reality: An Introduction to the Philosophy of Science*; Hansen, *New Philosophy for New Media*; Latour, *Politics of Nature: How to Bring the Sciences into Democracy*; Connolly, *Neuropolitics: Thinking, Culture, Speed*.
18. See B. H. Smith, *Contingencies of Value: Alternative Perspectives for Critical Theory*, esp. pp. 150–86, and *Belief and Resistance*, pp. 1–22.

Chapter 2

Pre-Post-Modern Relativism

If 'relativism' means anything at all, it means a great many things. It is certainly not, though often regarded that way, a one-line 'claim' or 'thesis': for example, 'man is the measure of all things', 'nothing is absolutely right or wrong', 'all opinions are equally valid', and so forth.[1] Nor is it, I think, a permanent feature of a fixed logical landscape, a single perilous chasm into which incautious thinkers from Protagoras' time to our own have 'slid' unawares or 'fallen' catastrophically. Indeed, it may be that relativism, at least in our own era, is nothing at all – a phantom position, a set of tenets without palpable adherents, an urban legend without certifiable occurrence but fearful report of which is circulated continuously. That would not mean, of course, that the idea was without consequence. On the contrary, no matter how protean or elusive relativism may be as a doctrine, it has evident power as a charge or anxiety, even in otherwise dissident intellectual quarters, even among theorists otherwise known for conceptual daring. It is this phenomenon that I mean to explore here: not relativism per se, if such exists, but the curious operations of its invocation in contemporary intellectual discourse and something of how it came to be that way.

The historical angle will be significant here, as indicated by my title, intended to evoke a relativism that is both 'pre-' and 'post-' modern but still also 'modern' – or, as I shall elaborate below, Modern*ist*. The point here is not so much that the views so named are perennial, though that, too, could be maintained. Given current understandings of the terms in question ('postmodern' as well as 'relativist'), one could claim as pre-'postmodern relativists' all those from Heraclitus and Montaigne to Alfred North Whitehead or Ludwig Wittgenstein who have questioned ideas of epistemic, moral or ontological fixity, unity, universality or transcendence and/or who have proposed correspondingly alternative ideas of variability, multiplicity, particularity or contingency. The point is, rather, that the periods of the emergence and prevalence of such views remain open questions for intellectual history and, therefore, that any presumed or asserted historical specificity is suspect.

Considerable recent work in intellectual history suggests that, from the end of the nineteenth century and increasingly to the eve of the Second World War, a notable feature of theory in virtually every field of study was a more or less radical questioning of traditional objectivist, absolutist and universalist concepts and a related effort to develop viable alternative – non-objectivist, non-absolutist, non-universalist – models and accounts.[2] Major representative figures involved in such activities, both critical and productive, include Friedrich Nietzsche, Martin Heidegger and John Dewey in philosophy; Ernst Mach, Albert Einstein and Niels Bohr in physics; Karl Mannheim in social theory; Franz Boas in anthropology; and Edward Sapir in linguistics. If 'relativism' is understood most generally and non-prejudicially as this sort of radical questioning and related theoretical production, then we may observe that, in the era we call 'Modernist', it appears to have been a significant strand in much respectable intellectual discourse. Stated thus, the observation may not be contentious. It is worth stressing, however, in view of the current routine attachment of the ostensible period-marker 'postmodern' to ideas also characterised as 'relativist' and the operation of that double label – 'postmodern relativism' – as the sign of a distinctly contemporary as well as especially profound intellectual, moral and political peril.

I

To begin to explore how such characterisations operate, we may consider a few journalistic examples. A recent review in the *New York Times* discusses two books concerned with the trial of scholar Deborah Lipstadt in a libel suit brought against her as author of a work titled *Denying the Holocaust*.[3] One of the books under review is by British historian Richard Evans, Lipstadt's key witness at the trial and himself the author of an earlier work described by the reviewer, Geoffrey Wheatcroft, as 'an attack on postmodernism and deconstructionism in the name of the traditional historical virtue of objectivity'.[4] The other book is by the relatively young American journalist D. D. Guttenplan, whose account of the trial Wheatcroft praises but whose 'ventures into theory' he describes as 'less happy'. The evidence of this infelicity is Guttenplan's rejection of the idea, put forward by Evans, of a link between Holocaust denials and 'an intellectual climate in which "scholars have increasingly denied that texts have any fixed meaning" '.[5] Wheatcroft remarks:

> But surely Evans's point is well taken precisely in this context. Once we allow the postmodernist notions that historical data are relative, that all truth is subjective and that one man's narrative is as good as another's, then Holocaust denial indeed becomes hard to deal with.[6]

Two features of the passage are worth noting. One is the utter invisibility of any nameable, citable, quotable proponents of that cascade of 'postmodernist notions'. The other is the hodgepodge quality of the notions themselves, which range from sophomoric slogans to important ideas currently at issue and by no means self-evidently absurd. Who among the figures commonly associated, properly or improperly, with 'postmodern' theory maintains that all truth is subjective or that one man's narrative is as good as another's? Michel Foucault? Jacques Derrida? Jean-François Lyotard? Hayden White? Richard Rorty? Stanley Fish? David Bloor? Bruno Latour? Actually, of course, none of these. Similarly, is it quite clear that texts *do* have fixed meanings and that historical data are *not* relative to anything – for example, to the perspectives from which they are viewed or to the idioms available for reporting them? The parading of such dependably – if not always relevantly or inherently – scandalising ideas and the absence of specific citations (authors, texts, passages) for any of them are standard features of the contemporary invocation/denunciation of 'postmodern relativism', whether done crudely, as here, or more artfully, as we shall see later, at the hands of scholars, academic critics and philosophers.

To continue, however, with our example: the idea of an atmospheric linkage between Holocaust denial and relativistic 'postmodern' theory – floated by Evans and endorsed by Wheatcroft – is central to Lipstadt's own book, subtitled 'The Growing Assault on Truth and Memory'. Explaining her conviction that 'part of the success' of current denials of the Holocaust 'can be traced to an intellectual climate that has made its mark on the scholarly world during the past two decades',[7] she continues:

> Because deconstructionism argued that experience was relative and nothing is fixed, it created an atmosphere of permissiveness toward questioning the meaning of historical events and made it hard for its proponents to assert that there was anything 'off limits' for this sceptical approach . . . No fact, no event, and no aspect of history has any fixed meaning or content. Any truth can be retold. Any fact can be recast.[8]

'This relativistic approach to the truth,' Lipstadt observes, 'has permeated the arena of popular culture, where there is an increased fascination with, and acceptance of, the irrational', an observation illustrated by belief in alien abduction.[9]

Several points can be made here. First, Lipstadt's conception of the operations of causality in intellectual history, both general and specific (what causes/caused what, how conditions for the emergence of certain ideas or claims arise/arose, and so forth), is exceedingly vague and otherwise dubious. No less dubious is her representation of scepticism as an inherently worrisome 'approach'. It is certainly not the arguments of deconstruction (such as they may be) or any consequent atmosphere of academic permissiveness (to the extent that such exists) that inspire Nazi-apologists to deny the systematic extermination of Jews in Germany. Nor is it deconstruction or academic pessimism that makes such denials credible among ill-educated segments of the population. Indeed, it could be argued that, if it *is* 'deconstructionism' or 'relativism' that leads historians and other members of the intellectual community to regard every received fact, truth and belief without exception as open to question, then we should be grateful that *something* in the atmosphere encourages critical reflection when so much else in it encourages dogmatism and self-righteousness. This is not to say that it is dogmatic to maintain that the events we call the Holocaust occurred. But it is certainly a recipe for dogmatism to maintain, as Lipstadt does here, that the 'meaning and content' of those events should be 'off limits' to redescription or reinterpretation. That particular events may be recast from deliberately malign perspectives is a risk that attends a communal ethos of openness to critical reflection and revision. But the risk of communal self-stultification created by the muzzling of scepticism – or by its attempted quarantine as a contagious moral ill – could be seen as greater and graver by far.

Elsewhere in her book, Lipstadt invokes relativism in another way that is recurrent and in some respects fundamental in contemporary usage: that is, as a deeply improper claim of equivalence, similarity, continuity or comparability between things that are clearly and unquestionably (or that is the crucial presumption in such cases) unequal, different, distinct and incomparable. Thus, referring to works by revisionist German historians who compare and stress similarities between the Holocaust and other massive state-sponsored slaughters, Lipstadt maintains that the 'relativist' historians in question 'lessen dramatic differences', 'obscure crucial contrasts' and produce 'immoral equivalences'.[10] It is proper, of course, for Lipstadt and other scholars to expose the limits of such comparisons, especially where their evident motive or effect is to minimise specific crimes or to exculpate specific agents or policies. But to denounce as 'immoral' the observation of similarities (contextual, procedural and so forth) between some specific event and all other events is to claim for the former an absolute uniqueness that not only attests to the impossibility of historical thought in that regard for oneself (understandable in the case of survivors and their families) but would bar such thought in that regard for everyone else.

The association of relativism with morally improper comparisons recurs in a newspaper column that appeared shortly after 9/11 under the arresting headline, 'Attacks on U.S. Challenge Postmodern True Believers'. According to the columnist, Edward Rothstein, the murderous attacks on American targets exposed the hollowness of 'postmodernist' – and here also 'postcolonialist' – relativism. He explains:

> [P]ostmodernists challenge assertions that truth and ethical judgment have any objective validity. Postcolonial theorists . . . [suggest] that the seemingly universalist principles of the West are ideological constructs . . . [and] that one culture, particularly the West, cannot reliably condemn another, that a form of relativism must rule.[11]

But, Rothstein continues, 'this destruction seems to cry out for a transcendent ethical perspective'. '[E]ven mild relativism' that 'focuses on the symmetries between violations' is 'troubling'; for what are 'essential now' are 'the differences . . . between democracies and absolutist societies' and also between 'different types of armed conflict', by which he presumably means something like inherently unjustified terrorism as distinct from manifestly just wars of defence.

The questions raised by this equation of relativism with morally improper comparisons are, as in Lipstadt's case, themselves ethical, involving here political as well as intellectual judgements and acts. Rothstein evidently sees no relation between what he denounces as the 'ethically perverse' idea of symmetry – which, he claims, requires a 'guilty passivity' in the face of manifest wrong[12] – and what he calls for as a 'transcendent ethical perspective'. But symmetry – that is, an observable correspondence between elements of otherwise different or opposed things and, accordingly, their equitable or proportional treatment – is closely related to common ideas of fairness and could be seen as a crucial aspect of justice.[13] Rothstein also sees no connection between the 'unqualified condemnations' he regards as necessary in this case and the 'absolutism' that, in his view, characterises societies so different from Western democracies that only a postmodern relativist could think of considering the two symmetrically. In the days immediately following 9/11, a number of regional specialists and other academic commentators urged consideration of the less obvious conditions plausibly involved in motivating the attacks, including what they saw as the relevant culpabilities of United States policies in the Middle East. All these public commentators, however, condemned the attacks per se. What Rothstein appears to mean by 'unqualified condemnation', then, is a refusal to accept any consideration as bearing on the judgement of certain matters and a refusal to acknowledge the desirability of any reflection on them. Here, as often elsewhere, a denunciation of relativism amounts to a demand for dogmatism – for predetermined judgement armoured against new thought.

II

The next example is considerably more sophisticated than those just examined and illustrates not the familiar charge of relativism but its no less familiar disavowal. The passage in question, from Donna Haraway's recent book *Modest_Witness@Second-Millennium,* concerns changing contemporary understandings of scientific knowledge. Referring to anthropologists' accounts of negotiations between the Australian government and Aborigine tribal representatives over land-surveying techniques, Haraway notes the increased recognition in the 'postcolonial world' of '"indigenous" ways of knowing'. She continues:

> Even more challenging to most Western ideas about knowledge, science itself is now widely regarded as indigenous and polycentric knowledge practice. That is natural science's strength, not its weakness. Such a claim is not about relativism, where all views and knowledges are somehow 'equal', but quite the opposite.[14]

Haraway's explanation of how this challenging new conception of science is 'quite the opposite' of relativism is curious, but instructive here along several lines. She writes:

> To see scientific knowledge as located and heterogeneous practice, which might (or might not) be 'global' or 'universal' in specific ways rooted in ongoing articulatory activities that are always potentially open to critical scrutiny from disparate perspectives, is to adopt the worldly stance of situated knowledges. Such knowledges are worth living for.[15]

What makes this explanation curious is that, aside from equivocal phrasing and punctuation (to which I return below), its elaboration of why such a view is *not* relativism would strike a number of Haraway's readers as exactly what they mean, for better or worse, by relativism. Here, for example, is an unapologetic relativist credo by historian of science Kenneth Caneva:

> I believe that a full understanding of *anything* entails an understanding of its history, of its process of becoming. Historical relativism – the relativism that

> refers to the historically and personally contingent connectedness of all human actions and beliefs – is about as undeniable a fact of our existence as I can imagine. . . [T]his way of looking at things does not make scientific knowledge no better than the opinion of many people . . . [The warrant of scientific knowledge] derives from a richly interconnected web of researchers, extending widely, both in space and time, and employing exacting (if historically developed) norms. [In] this kind of relativism . . . the things to which scientific knowledge is relative are broad and deep intellectual, social, and instrumental connections, repeatedly scrutinized, challenged, defended, and modified.[16]

As I think is clear, the elaborated conception of scientific knowledge that Caneva offers here in explicating his avowed position of relativism is virtually indistinguishable from the conception of such knowledge that Haraway offers as demonstrating why the view she commends is 'quite the opposite' of relativism.[17]

Virtually indistinguishable. There are, however, some differences, notably certain equivocating gestures in Haraway's account, such as her reference to 'practice . . . which might (or might not) be "global" or "universal" in specific ways' and, more generally, formulations that slide between, on the one hand, strenuous affirmations of such radically up-to-date ideas as the 'located'-ness and 'rootedness-in-ongoing-activities' of all knowledge practices and, on the other hand, nods toward more traditional views of such matters. Among the latter is the idea that what distinguishes genuine scientific knowledge from 'primitive' or 'pre-scientific' beliefs is precisely the fact that its validity is *not* rooted in local, culture-specific activities but, on the contrary, is transcultural or, as it is usually said (and as Haraway herself says, but in hedging quote-marks), 'universal'. To the extent that Haraway's formulations make concessions to such ideas, they underwrite her disavowal of the traditionalist's bugbear, relativism. To that extent, however, they also blunt the force of what she otherwise offers as a critique of just those traditional ideas and compromise the conceptual clarity, consistency and workability of the alternative views she proposes as their replacement – for example, what she calls in this passage the 'stance of situated knowledges' and elsewhere, no less problematically, 'strong objectivity'.[18]

If there is a puzzle in Haraway's recurrent identification of relativism with a fatuous egalitarianism[19] and pointed dissociation of herself from the acknowledged relativist positions of science-studies scholars whose views, in the crucial respects, she largely shares, the solution (or at least part of it) may be suggested by her remark, at the end of the second passage quoted above, that, as understood by the 'worldly stance of situated knowledges', such knowledges are 'worth living for'.

The remark, which might otherwise appear gratuitous, seems to signal Haraway's sensitivity to the current ideological burden of the charge of relativism in politically activist – for example, feminist and Marxist – circles. In particular, it seems to respond to the widespread conviction and anxiety in such circles (as, of course, elsewhere) that, without the possibility of appeals to objective, universal (that is, *non*-'situated') standards, one would be left in a state of nerveless lassitude – indifferent to what anybody does, unwilling to judge anything, unable to choose anything, and thus without anything 'worth living [or, presumably, dying] for'. Given the general canniness of Haraway's thought, there is good reason to think that she appreciates the strained logic and dubious view of human motivation involved in such convictions and related anxieties. That she nevertheless plays to both in her readership, rather than explicitly challenging them, may be a determined strategic decision on her part. But it is, I think, an unfortunate one in the long run, both intellectually and politically.[20]

III

For an instructive perspective on some of the issues raised above, including the nature of historical truth and the relation of science to other knowledge practices, I turn now to the Modernist part of this story, beginning with two important and in some ways exemplary figures of the period. One is the American intellectual historian, Carl Becker; the other is the Polish microbiologist and historian-sociologist of science, Ludwik Fleck. The relevant – which is to say, relativist – works of both appeared in the 1930s.

Becker, a cosmopolitan Midwesterner widely celebrated for his study, *The Heavenly City of the Eighteenth-Century Philosophers*, was elected president of the American Historian Association in 1932. In his presidential address that year, titled 'Everyman His Own Historian', he elaborated the idea that the activities of the professional historian are not fundamentally different from the sorts of trace collecting, trace interpreting and narrative construction performed by laypeople in regard to personal, family and local histories. For Becker, this continuity of formal and informal historiography implied, among other things, that, contrary to the positivism then dominating the profession, historians should not take the work of natural scientists as their model of intellectual activity.

In the same years that Becker was chiding his fellow historians for their misplaced identification with scientists, Fleck – himself a practising scientist – was arguing that science is fundamentally continuous with everyday knowledge construction. In his major work, *Genesis and Development of a Scientific Fact*, Fleck challenged the prevailing idea that there were specific features of genuinely scientific knowledge (systematic testing, empirical verifiability, practical applicability and so forth) that marked it off clearly both from primitive belief and from the errors of the less scientifically enlightened past. In the course of his analysis and critique of such views, which he censured as historically short-sighted and intellectually confining, he observed:

> Whatever is known has always seemed systematic, proven, applicable, and evident to the knower. Every alien system of knowledge has likewise seemed contradictory, unproven, inapplicable, fanciful, or mystical. May not the time have come to assume a less egocentric, more general point of view? [21]

Fleck's name for such an empirically broad-based, non-self-flattering study of knowledge was 'comparative epistemology'. His elaboration and illustration of it in *Genesis and Development* figured crucially in the development of Kuhnian and post-Kuhnian (or 'postmodernist') sociology and history of science.[22] For a number of current mainstream philosophers, it figures as the very model of 'extreme' epistemic relativism.[23]

For Fleck, the emergence and specific features of what we experience as 'fact', 'truth' or 'reality' are made possible, but also severely constrained, by the systems of ideas, assumptions and related perceptual and classificatory dispositions (or, in his phrase, 'thought styles') that prevail in our particular epistemic communities – that is, disciplines, fields of specialisation, sub-cultures and so forth, or what he called 'thought collectives'. Tracing a significant tradition in the social study of knowledge from Comte to Durkheim, Fleck criticises, extends and radicalises the thought of these already quite innovative theorists: 'All these sociologically and humanistically trained thinkers . . . no matter how productive their ideas, commit a characteristic error: they exhibit an excessive respect, bordering on religious reverence, for the facts of natural science.'[24]

Commenting on the views of Lucien Lévy-Bruhl, who contrasts 'the mentality of primitive societies', which do not yet have 'a feeling for, or knowledge of, what physically is possible or impossible', with scientific concepts, which 'solely express objective features and conditions of beings and phenomena',[25] Fleck writes:

> We must object in principle that nobody has either a feeling for or knowledge of what physically is possible or impossible. What we feel to be an impossibility is merely an incongruence with our habitual thought style . . . 'Experience as such' . . . is chimerical . . . Present experiences are linked with past ones,

thereby changing the conditions of future ones. So every being gains 'experience' in the sense that he adjusts his way of reacting during his lifetime.[26]

Thus, as Fleck stresses, the scientist's perceptions of the physical world are no more objective than those of anyone else since, like anyone else's, they are shaped by a particular experiential history in a particular social-epistemic community.[27]

It is unlikely that Fleck or Becker knew each other's writings, but their intellectual affinities are evident. These include a shared interest in and extensive familiarity with the broader intellectual and cultural worlds in which they worked and lived. Thus Fleck, a microbiologist by profession and medical historian by avocation, read widely in early twentieth-century anthropology, sociology and psychology and was well acquainted with the philosophy of science of his time, including the work of the Vienna Circle.[28] Becker, comparably, was both a highly respected specialist in his field and immensely literate. Seeking to account for his unorthodox and irreverent views, later commentators point to his temperamental affinities with the Enlightenment ironists who were his subjects (Pascal, Diderot, Voltaire and so forth) but also note Becker's willingness, as one of them puts it, 'to play hooky from his work, reading the fiction of . . . James Joyce, Anatole France, Henry James, Virginia Woolf, and Proust'.[29] Whatever else might be said about the relativistic character of their thought, it was certainly not provincial.

Becker and Fleck were both also explicitly self-reflexive. Thus Becker writes at the conclusion of 'Everyman His Own Historian':

I do not present this view of history as one that is stable and must prevail. Whatever validity it may claim, it is certain, on its own premises, to be supplanted . . . However accurately we may determine the 'facts' of history, the facts themselves and our interpretations of them, and our interpretation of our own interpretations, will be seen in a different perspective . . . as mankind moves into the unknown future. Regarded historically, as a process of becoming, man and his world can obviously only be understood tentatively, since it is by definition something still in the making, something as yet unfinished.[30]

Similarly, Fleck stresses the need for historical assessments of all scientific claims, including those of one's own era. Referring to the teachings of eighteenth-century medical pathology (for example, the 'humour' theory of illness), he writes:

It is perfectly natural that these precepts should be subject to continual change . . . It is altogether unwise to proclaim any such stylized viewpoint, acknowledged and used to advantage by an entire thought collective, as 'truth or error'. Certain views advanced knowledge and gave satisfaction.

> These were overtaken not because they were wrong but because thought develops. Nor will our opinions last forever, because there is probably no end to the . . . development of knowledge just as there is probably no limit to the development of other biological forms.[31]

The point – that is, the explicit reflexivity of both Fleck and Becker – needs emphasis in view of the routine charge that relativists, in (assumedly) claiming absolute truth for their own relativist views, are self-refuting.[32]

A final note may be added on these two representative figures of Modernist relativism. In an essay titled 'What is Historiography?', Becker observes that a properly historical understanding of 'history' would see it not as 'the gradual emergence of historical truth objectively considered' but, rather, as

> the gradual extension of this time and space world . . . the items, whether true or false, which acquired knowledge and accepted belief enabled men (and not historians only) to find within it, and the influence of this pattern of true or imagined events upon the development of human conduct . . . Nor would the historian be more interested in true than false ideas about the past.[33]

One may be struck by this last point, that is, Becker's suggestion that false ideas require the attention of historians no less than true ones; for it would be, fifty years later, a key methodological plank – the well known 'symmetry principle' – in the Edinburgh-based Strong Programme in the history and sociology of scientific knowledge.[34] Of particular interest here, however, is the point that precedes it, that is, Becker's suggestion that our 'time and space world' is extended as a 'pattern of true or imagined events' that we 'find' – or, as we would say now, 'construct' – through 'acquired knowledge and accepted belief'. The idea is complex and somewhat awkwardly framed here, but clearly attuned to the constructivist views of knowledge formulated by Fleck and other social theorists at the time and developed throughout the twentieth century by various historians and sociologists of science, theorists of learning, perception and cognition, and a few dissident philosophers of science.[35] As indicated in Chapter 1, it is this constructivist-interactionist-pragmatist account of cognitive dynamics that operates today as the major rival to the realist-rationalist-representationalist understandings of cognition that continue to dominate formal epistemology, analytic philosophy of mind and mainstream philosophy of science. And it is the alternative conceptualisations of truth, fact, science, objectivity and so forth emerging from this constructivist tradition in historiography and science studies that are the object of some of the most strenuous expressions of outrage at 'postmodern relativism'.[36]

IV

If there is nothing especially new about the views now characterised as 'postmodern' relativism, there is even less new about the modes and occasions of their denunciation and disavowal. An instructive Modernist example of the latter occurs in Proust's *Time Regained*, where the narrator, reflecting on the subjectivity of our perceptions of other races as illustrated by the 'faculty [of Germanophiles] of ceasing for a moment to understand and even to listen when one spoke to them about the German atrocities in Belgium', adds in parenthesis:

> yet they were real, these atrocities: the subjective element that I had observed . . . did not imply that an object could not possess real qualities or defects and in no way tended to make reality vanish into pure relativism.[37]

The disavowal here is interesting but somewhat paradoxical. On the one hand, it indicates that, in Proust's time as in our own, relativism was readily associated with a denial of objective reality and, as such, seen as especially morally culpable when linked to well-known atrocities. At the same time, however, the narrator's fascination – in this passage as throughout *Remembrance of Things Past* – with multiple perspectives and contingent perceptions and the significance of the effects of such psychological phenomena for the structure and major themes of the novel align Proust's thought with the relativistic currents of his era being noted here.

By the 1920s and 1930s such currents were strong enough to create alarm in the philosophical community and to elicit efforts at formal rebuttal. One of the most sustained anti-relativist efforts of the period is Maurice Mandelbaum's *The Problem of Historical Knowledge: An Answer to Relativism*. Mandelbaum's argument is, first, that the claim of objectivity by scientists and historians is crucial to their authority and, second, that the claim can withstand the scepticism of critics and theorists when avowed conscientiously and attended by various self-effacing methods.[38] Here as later, the 'refutation' of relativism consists largely of a reaffirmation of the conventional ideas at issue and the defence of those ideas consists largely of a rehearsal of the reasons conventionally given in support of them.

To a practising historian such as Becker, it was obvious that what one finds in the archives and how one reports and assesses those findings will be affected by one's purposes, concerns and perspectives. Expression of such a view, however, was commonly interpreted as an assertion of the bias of all professional historians and thereby as a slander on their dignity. Also, by a familiar logic, such views were thought to imply that all accounts of a historical event are equally valid and thus (by the logic in play) equally worthless. These overheated interpretations and gratuitous inferences recur in virtually the same terms in late twentieth-century responses to the supposed assaults by 'postmodernists' on truth, reason and 'the traditional historical virtue of objectivity' – that is, to efforts by scholars and theorists in fields such as historiography, literary studies and the sociology of knowledge to indicate the historicity of such terms as *fact*, *truth* and *objective* and to explore their ideological and institutional operations.[39] The extent of the duplication can be startling and/or, depending on one's point of view, amusing. For example, in an especially grim anti-relativist tract of the period titled 'The Insurgence against Reason', American philosopher Morris Cohen maintains that the '[present] decline of respect for truth in public . . . affairs is not devoid of all significant connection' to 'the systematic scorn heaped by modernistic [sic] philosophies on the old ideal of the pursuit of truth for its own sake'.[40] And this was 1925.

Then as now, challenges to received views were labelled 'irrationalist' and linked to other egregious if not always well understood contemporary disturbances, such as 'Bolshevism' and relativity theory.[41] Then as now, worrisome correlates or consequences of such challenges were discerned in popular culture and belief (one recalls Lipstadt's association of 'deconstructionism' with belief in alien abduction), though, in the Modernist period, they were seen also in the 'deformations' of avant-garde art, music and literature. Einstein and Lenin, Woolf and Stravinsky, Joyce and Picasso: pre-postmodern relativists all!

V

Regarded retrospectively, the early twentieth-century critiques of objectivist, absolutist, universalist assumptions in historiography and epistemology, along with the elaboration of alternative concepts, models and accounts in those and related fields, were exceptionally intellectually fertile. Indeed, some of the most significant developments of later twentieth-century thought, including the new historiography of the *Annales* school, constructivist epistemology, post-Kuhnian history and sociology of science, and poststructuralist language theory, could be seen as extensions and refinements of those early critiques and alternative 'relativist' accounts. The claim becomes stronger, though ironically so, if we note that two other important movements in twentieth-century thought, analytic philosophy and Frankfurt School Critical Theory, operated to a considerable extent as conservative reactions to those developments and that the central aims and issues of contemporary academic philosophy have been shaped accordingly. Thus much mainstream philosophical activity over the past thirty or forty years has consisted of efforts either to shore up the objectivist-universalist views still at issue or to discredit the alternative views still being elaborated.[42] And, of course, vice versa. That is, a good bit of the energy of theoretical radicalism over the same period has consisted of rebutting purported exposures of logical absurdity, answering charges of moral quietism or political complicity, or attempting to escape the fray by artful navigations.[43]

The subsequent fortunes of the Modernist/relativist views evoked above have been quite variable, reflecting, among other things, shifting intellectual moods in the academy and elsewhere that have themselves been responsive to broader social and political events marking this past very eventful century. Especially significant here are the European and North American experience of the Second World War; the effects of the hyper-reactive McCarthy era in the United States; the amalgam of political activities, popular beliefs and cultural representations that made up the Cold War; the global eruption of various radical social movements (anti-colonialist, civil rights and Black Power, feminist, anti-war, counterculture and so forth); and, throughout the century, dramatic technological developments and widespread demographic shifts. The effects of such events, trends and developments on later twentieth-century intellectual life are too complex and complexly mediated to be traced here, but we might recall briefly some of the most relevant moods and movements.

Although the interwar period from the 1920s to the late 1930s was marked by a confident positivism in the natural sciences and a related scientism in much academic philosophy, there was, in other quarters of the academy as in the earlier years of the century, a continued radical questioning of positivist, realist and universalist views. With Franz Boas and his students, including Margaret Mead and Ruth Benedict, cultural relativism became a respectable if never wholly dominant view in anthropology. In economic theory and political science, there was an increased emphasis on the irreducibility and partiality of subjective perspectives, with invocations of non-Euclidean geometry and relativity theory in physics as pertinent to the understanding of social phenomena.[44] In linguistics, a number of influential theorists, among them Edward Sapir and Alfred Korzybski, explored the cultural variability and ideological power of the operations of language. In American philosophy, there were of course the pragmatists, notably, in this period, John Dewey; and, in historiography, Charles Beard as well as Carl Becker continued to challenge the positivist ideals and objectivist claims of their fellow historians. Commenting on the affinities among some of these movements in this period, Peter Novick writes:

> Pragmatism's crusade against the worship of facts, its scepticism about claims of objectivity, its consistent reluctance to accept a hard and fast fact-value distinction, its emphasis on change and flux, on the human and social elements in knowledge, and stress on the practical consequences of knowledge – all these were at the center of the relativist sensibility.[45]

The post-war years from 1945 to the 1950s were a period of anxious social conservatism in the United States and, in the academic and intellectual world, a time of pulling back from radical theory, especially from historicist and relativist directions. Thus, in linguistics, the anthropological approaches of Sapir and Benjamin Lee Whorf contended with increasingly aggressive turns (and returns) to universalist and rationalist accounts, a development that reached an apogee of sorts in the late 1950s with the arrival and rapid, widespread embrace of Chomsky's transformational-generative linguistics. In literary studies, the old positivist historical philology and I. A. Richards' proto-reception-theory were both upstaged by an intensely formalist, explicitly anti-historicist New Criticism. (W. K. Wimsatt, Jr, promoting what he called 'objective criticism' in 1946, deplored Richards' *Principles of Criticism* [1929] for committing the so-called Affective Fallacy: 'a confusion between the poem and its *results*' that is 'a special case of epistemological scepticism . . . and ends in impressionism and relativism'.[46]) Among professional historians, there was a renewed (war- and propaganda-chastened) commitment to the idea and ideals of objectivity and, in departments of philosophy, a withdrawal from the capacious concerns and diverse approaches of the 1930s and 1940s (aesthetics, ethics, philosophy of education, political theory, phenomenology, existentialism and even some empirical, activist and popular ventures into politics and education) to the more confined pursuits and technical, formal, logicist methods that became known as 'analytic philosophy'.[47]

For many academics, especially younger ones, the mood shifted significantly again in the late 1960s and 1970s, which saw an irruption of social and political radicalism in Europe and the United States and, with it, a self-conscious 'revolutionising' of theory in a number of fields.[48] In epistemology and the history and philosophy of science, it was exemplified most visibly by Kuhn's *Structure of Scientific Revolutions* (1963), the self-declared relativisms of Paul Feyerabend's *Against Method* (1975) and Nelson Goodman's *Ways of Worldmaking* (1978), and Richard Rorty's anti-foundationalist treatise, *Philosophy and the Mirror of Nature* (1979). These were paralleled in language theory, literary and cultural studies, and historiography by the appearance or importation of, first, structuralism and semiotics, followed soon after by Derridean deconstruction, poststructuralism, Foucauldian new historicism, and the studies that popularised, with its myriad meanings, the term 'postmodern'.[49] Conservative responses to each of these developments were not lacking from the philosophical community, among the most influential of which were the presumptive demolition of Kuhn by Israel Scheffler, the presumptive disposal of Kuhn, Feyerabend and Whorf by Donald Davidson, the rebuffs of 'postmodernism' by Jürgen Habermas, and the work of a number of other denouncers and alleged devastators of relativism in a widely cited volume, *Rationality and Relativism*, published in 1982.[50]

The past twenty-five years have witnessed a number of major disintegrations, migrations and realignments – and, with them, more or less violent local antagonisms – in the social and political sphere and, comparably and relatedly in the intellectual and academic worlds, a situation of increasing ideological multiplicity, heterogeneity, shift, clash and conflict. Thus we have the emergence of the multicultural university, cultural studies, identity politics and the associated 'culture wars'; revisionist history and historiography, both left-wing and right-wing, and the associated 'history wars'; post-Kuhnian science studies, constructivist-pragmatist epistemology and the associated 'science wars'; and, finally, the continued playing out of poststructuralist thought in the humanities and social sciences, and the associated 'theory wars'. In all these not-altogether-academic conflicts, the charge, fear or denunciation of relativism has operated with egregious frequency, if not always obvious relevance.

Two points may be stressed here with regard to the contemporary scene. One is the institutional co-presence of, on the one hand, scholars who entered their disciplines during the 1950s and early 1960s and remain committed to the projects and methods that prevailed at that time (and also persuaded by the traditional justifications for each) and, on the other hand, scholars trained in, comfortable with, and seeking to pursue the sorts of approaches ('continental', poststructuralist, new historicist, feminist, postcolonialist, and so forth) that emerged in the succeeding decades. Indeed, most of the 'wars' mentioned above resolve into de facto generational struggles, even if not always between literally younger and older scholars. What is also significant about this last phase and particularly the series of conflicts just mentioned is their creation of a widespread demand for munitions and arms-suppliers and, thereby, the conditions for the emergence of a large scale anti-postmodern-relativism industry, with branch factories in virtually every field of study. Thus we have a string of academic publications over the past twenty years with titles such as *Looking into the Abyss*, *The Truth about Truth*, *The Truth about Postmodernism*, *The Illusions of Postmodernism*, and so forth; the ongoing labour of relativism-refutation by old and new generations of academic philosophers; and the invocation and rehearsal of the presumptive triumphs of all these by students and colleagues, fans and followers, citers and re-citers.

VI

As a final example here, we may consider a recent book that draws on the extensive arsenal just mentioned to mount a frontal assault on 'postmodern relativism' in the literary academy. The book, by self-declared left-wing literary scholar Satya Mohanty, is titled *Literary Theory and the Claims of History: Postmodernism, Objectivity, Multicultural Politics*. Mohanty's political self-identification is relevant here because his explicit aim is to persuade his similarly left-wing colleagues and students ('feminist, antiracist, postcolonialist, and so forth') that the 'postmodernist' ideas many of them admire lead to an 'extreme relativism' that is, he writes, an 'untenable – and indeed rather dangerous – philosophical ally for political criticism'.[51] As Mohanty sees it, contemporary literary/cultural theory is beset by a debilitating scepticism about the possibility of rational argument and objective knowledge that would be relieved by better acquaintance with the accounts of knowledge and language developed some years back by a generation of now senior analytic philosophers.[52]

Noting the common dismissal of relativism as transparently absurd (he rehearses the argument from self-refutation), Mohanty observes that, because it does not 'operate in literary critical circles . . . as an explicit and reasoned position', a treatment of that sort is not available in this case. Rather, he continues, the relativism that concerns him is 'embedded in our critical gestures, in the kinds of questions we ask or refuse to ask' or implied by certain 'emphases and absences of emphasis' in the work of various poststructuralist critics and theorists.[53] A set of three passages, presented in tandem, are said to be representative.[54] The first passage, from the opening pages of Foucault's *Archaeology of Knowledge*, describes the increased attention paid by contemporary historians to discontinuities in intellectual history. The second, from Hélène Cixous's 'Laugh of the Medusa', is about the experience of writing expressly as a woman. The third, by post-Marxist theorists Ernesto Laclau and Chantal Mouffe, maintains that the idea of a monolithic working class is irrelevant for contemporary politics and political theory. According to Mohanty, the three passages illustrate an obsession with 'difference' and a concomitant failure to 'raise the issue of the possibility of commonalities' that, together, implicate poststructuralism in relativism and disqualify it as a morally and politically adequate response to contemporary cultural diversity.[55]

As Mohanty goes on to describe it, this embedded, implicit relativism is characterised by two further features. One is the conviction that 'it is necessary to conceive the Other as a radically separable and separate entity in order to command our respect'; the second ('the most extreme relativist formulation of the problem') is the claim that 'there are no common terms between and among different cultures'.[56] This combination of a conviction that nobody seems to harbour and a claim that nobody seems to have made (certainly not, for either of them, the writers and theorists specifically cited as representative) constitutes the position of 'postmodern relativism', the exceedingly unhappy 'presuppositions' and 'entailments' of which Mohanty sets about to 'indicate' and 'clarify'. The process consists of the deployment of a chain of questionable identifications and strained inferences, including the idea, familiar from the long line of anti-relativist tracts and treatises mentioned above, that a rejection of traditional notions of objective criteria implies that one cannot or should not judge anything. From this dubious inference, Mohanty goes on to extract the similarly familiar and supposedly definitively relativistic claim that everything is equally good, or, as he frames it in his idiom:

> To believe that . . . there can be no responsible way in which I can adjudicate between your space and my space by developing a set of general criteria that can have interpretive validity in both contexts . . . is to assert that all spaces are equivalent, that they have equal value.[57]

From this rather artifactual 'assert[ion]' Mohanty extracts the no less artifactual implication that 'I cannot – and consequently need not – think about how your spaces might impinge on mine or how my history is defined together with yours', from which he moves to the triumphantly dismal conclusion that, 'if that is the case, . . . then I need not take you seriously'.[58] Thus we arrive at the position or condition of neo-colonialist insouciance against which Mohanty would warn the politically well-meaning but inadequately philosophically trained fans of poststructuralist theory. As I think is clear, the 'extreme relativism' that 'entails' this chilly attitude toward the Other is altogether spurious. Maintained by nobody, it is a phantom position spawned by tendentious reading and a string of question-begging identifications, distinctions and oppositions generated by the very ideas at issue – that is, by the traditional rationalist-realist-referentialist views of knowledge and language that have been challenged by a wide range of theorists, Anglo-American as well as European, over the course of the past century and defended over the same period by the mainstream community of academic philosophers.

Of particular interest here in connection with the question of scientific knowledge (encountered above in the discussions of Haraway, Caneva and Fleck) is Mohanty's rehearsal of certain philosophical arguments involving the presumption of a universal minimal rationality and universal set of shared true beliefs. The arguments, which appeared separately in two essays that are often cited as decisively relativism-refuting, were developed by the prominent analytic philosophers, Donald Davidson and Charles Taylor.[59]

In Davidson's version, the posited obligatory principle of 'charity' – whereby, when confronted with otherwise unknown people who speak a different language from ours, we must, to get started understanding them, 'assume general agreement on beliefs'[60] – is designed to refute the claims of absolute incommensurability or total non-inter-translatability that Davidson sees lodged in statements by Kuhn, Feyerabend, Whorf and other heterodox mid-century theorists of science, knowledge and language. The theorists in question note, among other things, that the meanings of particular scientific (and other) terms and concepts depend on their operations in particular discursive-conceptual systems (for example, Kuhn's scientific paradigms or Whorf's American Indian or Indo-European languages) that are culturally and historically variable and may differ from each other more or less significantly. None of these theorists maintains that this systemic-semantic interdependence implies local or linguistic solipsism – that is, the view that we are locked into our own cultures or languages, totally and permanently cut off from understanding the ideas of people in other times or places or from communicating with those who speak different languages. That, however, is the bleak implication that Davidson draws from a highly select, querulously interpreted set of their statements[61] and, following him, that Mohanty draws by a comparable process from 'poststructuralism'. And it is that bleak – and arguably totally straw – implication that is refuted by the idea of a universal faculty of reasoning by virtue of which every human being is fundamentally equipped to appreciate and evaluate the thinking of every other human being and all concepts can be communicated inter-linguistically, cross-culturally and transhistorically, at least in principle.

This reassuring argument proves as handy today for Mohanty, seeking to refute the purportedly 'embedded', 'entailed' pessimistic, quietistic relativism of Foucault, Cixous and Spivak,[62] as it proved thirty years ago for Davidson, seeking to refute the purportedly implied pessimistic, solipsistic relativism of Kuhn, Whorf and Feyerabend. Never mind that the argument is circular, that the only evidence of a difference-dissolving universal faculty of reason is its being presupposed by the universalist claims at issue. Never mind that the only thing that anyone – Kuhn, Whorf, Foucault or Spivak – is/was especially pessimistic about is/was the conceptual and practical viability of the traditional realist-rationalist-referentialist accounts of knowledge and language that remain foundational for mainstream academic philosophy – and foundational also for the arguments that are said to 'refute' alternative accounts.

The idea of a universal minimal rationality was invoked a few years later by Charles Taylor, this time in order to refute the relativist views of British philosopher, Peter Winch. Winch, responding to the reflections of ethnographer Evans-Pritchard, suggested that the irrational-seeming beliefs and practices of people from other cultures (the standard examples are Zande ideas and rituals involving 'witches') might make sense when interpreted in relation to their own cosmologies and related systems of norms.[63] Taylor and, echoing him, Mohanty object that, while such a view may appear amiable and tolerant, it actually entails a patronising, sentimental and morally disabling conception of the Other. For, as Mohanty explains it, as long as we acknowledge a universal human capacity for rational agency and (no small matter) a common 'world' with 'causal features' amenable to objectively true or false descriptions, then it follows that there are transcendent norms in accord with which the Other's beliefs and practices are not just culturally different from ours but objectively false and wrong.[64] And, conversely, if we reject the idea of such a universal capacity, then it follows that we have contempt for the Other.

Mohanty goes on to outline, supposedly contra the epistemological scepticism and neo-colonial quietism of 'postmodern relativism', a conception of 'cross-cultural negotiation' and resolution of intellectual disagreements as 'complex form[s] of cooperative social practice', mutually respectful, potentially democratic and 'open to revision and change'.[65] Nothing in such a conception, however, at least so stated, is excluded by the views Mohanty indicates as 'postmodernist', and a number of such (more or less genial and vague) possibilities are formulated in terms hardly distinguishable from his own by theorists he specifically cites as 'postmodern' or 'relativist'.[66] Indeed, the forms of ethical, intellectual and political interaction opened up by the views in question and articulated at length by the theorists thus labelled are characteristically innovative, energetic, responsive, responsible and nothing if not respectful of 'the Other'.[67] Two further points, however, must be made. First, the proviso above ('at least so stated') is required because

Mohanty's views on proper cross-cultural negotiation are often framed in quite different terms that would give considerable pause to such theorists and to many other readers. For example, he maintains, again following Taylor, that, in order to exhibit our genuine, non-patronising respect for Others, we must be prepared to point out to them the objective falseness of their beliefs and objective wrongness of their practices – presumably (he gives no examples) beliefs about things like witches (and angels?) and practices such as female genital mutilation (and male circumcision?). The second point is that there *is* something in Mohanty's conception of effective cross-cultural communication that is excluded by the views he calls postmodern, namely, the conviction that such desirable forms of 'multicultural politics' could take place only under the objectivist, universalist, realist-referentialist-rationalist accounts of knowledge, language and human agency that are, for Mohanty, intellectual bedrock.

Here as elsewhere, a major motive in the supposed exposure of the perils of postmodern relativism is a commitment to some set of orthodoxies, often, as in Mohanty's case, orthodoxies seen as necessary for certain generally accepted goods – communication, cooperation, justice, justification and so forth. Accordingly, it is worth emphasising that, when the theoretically innovative works thus characterised are dismissed out of hand or reduced to sophomoric inanities, among much else that is lost are the elaborated, alternative accounts they offer of just such matters, including accounts and demonstrations of how judgement, justification and communication can and do occur in domains such as law, politics and education *in the absence* of the transcendent universals and extensive human commonalities that they are classically said to presuppose and *in the midst* of the contingent particulars and human differences that are classically said to make them impossible.[68]

VII

The spurious inference of solipsistic, quietist, pessimistic, fatuously egalitarian and otherwise scandalising or depressing 'claims' from the ongoing questioning and reconceptualisation of traditional views of knowledge, language and human agency issues in the phantom position of postmodern relativism. The charge of 'postmodern relativism', so constructed, when directed at those pursuing unorthodox lines of thought in epistemology, ethics and social and political theory, operates by the same rhetorical and institutional mechanisms as did the charges of cynicism, materialism and nihilism directed in the past (and, in some places, still) at secular, naturalistic challenges to received theological-humanistic accounts of the human. A scapegoat label like these, it seeks to dishonour, exorcise and exile from the community a potentially disruptive intellectual force. A tin of straw herrings, the readily dismissible 'notions' and 'theses' of the always-already-refuted relativist obscure the not so readily dismissible critiques and alternatives of genuine critics and rivals. By linking intellectually challenging views with moral disablement or political complicity, such labels and charges, now as ever, obviate the need among those otherwise positioned to do so – teachers, journalists, philosophers, scientists, and political activists – to confront the actual content of those views, to engage them intellectually and to present them fairly and fully to students, colleagues, followers and the public. And where such charges, labels and linkages come to dominate thought and discourse in some community, they also obviate the possibility of such confrontation and engagement by later generations. Thus is communal self-stultification instituted and perpetuated.

Accordingly, I suggest that, insofar as we see ourselves as intellectually responsible scholars and teachers, we banish the phantom 'postmodern relativist' from our dreams and nightmares, excise the scapegoat label from our lexicons, expose the straw 'claims' and 'theses' as what they are, and take on the task of engaging actually existing ideas, relativistic and other. That means engaging them capaciously, in their extended textual forms, concept by concept, analysis by analysis, argument by argument, example by example. And, insofar as we see ourselves as moral and political agents, I suggest that we accept the task of operating in the world in accord with our most profound convictions and values, and with all the consciousness we can muster of the limits of our knowledge, our sagacity and our righteousness.

Notes

1. I cite here some familiar past identifications and current usages. As is clear, they are not synonymous or mutually entailed. At the end of the nineteenth century, 'relativism' could be understood as '[the doctrine that] nothing exists except in relation' (*Oxford English Dictionary*). In 2001, it could be identified blithely and without example or citation as 'the doctrine that all views are equally good' (Nozick, *Philosophical Explanations*, p. 21). For the multiplicity of meanings operating in contemporary academic philosophy, see Harré and Krausz, *Varieties of Relativism*. Harré and Krausz identify, define, distinguish and assess with due (that is, predictably conservative) philosophical judiciousness a dozen or more such varieties, for example, 'moral relativism', 'epistemic relativism' and 'ontological relativism', each with its 'anti-objectivist', 'anti-universalist' and 'anti-absolutist' variants and each of those with its 'strong' and 'weak' or 'moderate' and 'extreme' versions.

2. For relevant recent studies, see Novick, *That Noble Dream*; Vargish and Mook, *Inside Modernism: Relativity Theory, Cubism, Narrative*; Pippin, *Modernism as a Philosophical Problem: On the Dissatisfactions of European High Culture*; Burrow, *The Crisis of Reason: European Thought, 1848–1914*; Friedman, *A Parting of the Ways: Carnap, Cassirer, and Heidegger*; Menand, *The Metaphysical Club: A Story of Ideas in America*, esp. pp. 337–408; and Herbert, *Victorian Relativity: Radical Thought and Scientific Discovery*.
3. Lipstadt, *Denying the Holocaust: The Growing Assault on Truth and Memory*. The suit was brought by the British Nazi-apologist historian, David Irving, who lost the case roundly.
4. Wheatcroft, 'Bearing False Witness'; Evans, *In Defense of History*.
5. Wheatcroft, p. 13, quoting Evans, *In Defense*.
6. Ibid.
7. Lipstadt, *Denying*, pp. 17–18. This malign climate is exemplified again by the idea, evidently absurd for Lipstadt as for Evans and Wheatcroft, that 'texts have no fixed meaning', illustrated here with brief statements by Richard Rorty and Stanley Fish. Lipstadt gives Novick, *That Noble Dream*, as her source but abbreviates Novick's duly extensive citations and omits his duly clarifying contextualisations (see Novick, p. 540).
8. Lipstadt, *Denying*, pp. 18–19.
9. Ibid., pp. 17–19.
10. Ibid., p. 215.
11. Rothstein, 'Attacks on U.S.'.
12. Ibid.
13. See B. H. Smith, *Belief and Resistance*, pp. 7–8.
14. Haraway, *Modest_Witness*, p. 137.
15. Ibid., pp. 138–9.
16. Caneva, 'Objectivity, Relativism and the Individual', pp. 330–1.
17. Caneva's insistence that his avowedly relativist position does not mean he thinks scientific knowledge is no better than anyone's opinion could be seen as paralleling Haraway's insistence that the closely comparable view she endorses is 'not relativism, where everything is somehow "equal"'. The difference is that, whereas Caneva accepts the propriety of the label and rejects its identification with an absurd position, Haraway rejects the label while reinforcing its identification with an absurd position. See also Note 19 below.
18. See Haraway, *Simians, Cyborgs and Women: The Reinvention of Nature*, p. 187.
19. Elsewhere Haraway makes the identification explicit and, accordingly, denounces relativism outright:

    > Relativism is a way of being nowhere while claiming to be everywhere equally. The 'equality' of positioning is a denial of responsibility and critical enquiry. Relativism is the perfect mirror twin of totalization in ideologies of objectivity; both deny the stakes in location, embodiment, and partial perspective; both make it impossible to see well. Relativism and totalization are both 'god-tricks' promising vision from everywhere and nowhere equally and fully, common myths in rhetorics

> surrounding Science. But it is precisely in the politics and epistemology of partial perspectives that the possibility of sustained, rational, objective enquiry rests. (Haraway, *Simians*, p. 191)

The identity of those (or anyone) making such claims or promises and playing such tricks remains mysterious.

20. For related discussion, see Chapter 4.
21. Ludwik Fleck, *Genesis and Development of a Scientific Fact*, p. 23.
22. See Chapter 3 for details of Fleck's influence.
23. See Harré and Krausz, *Varieties of Relativism*, pp. 75, 100, 112–13.
24. Fleck, *Genesis*, p. 47; translation modified.
25. Claude Lévy-Bruhl, quoted in Fleck, *Genesis*, p. 48.
26. Fleck, *Genesis*, p. 48.
27. The received view, which Fleck disputes here, was that from its beginnings in primitive mentality eventually a higher type of cognition develops, in which – as manifested most notably in the European Enlightenment and Western science – thought is freed from custom and superstition, purified of subjective bias and emotion, and finally becomes capable of observing and describing the objective properties of things. Fleck's objections to this view challenge empiricist-rationalist understandings of human perception that still prevail in mainstream philosophy of mind and related theory in cognitive science. The accounts of perceptual and cognitive processes that he offers in their stead are in line with the work of mid-century developmental and experimental psychologists and anticipate later developments in cognitive neuroscience (see, for example, Piaget, *The Construction of Reality in the Child*; Gibson, *The Senses Considered as Perceptual Systems*; von Glasersfeld, *Radical Constructivism*; Edelman, *The Remembered Present*; Purves and Lotto, *Why We See What We Do: An Empirical Theory of Vision*).
28. See the bibliography in Fleck, *Genesis*, pp. 169–91. On the intellectual contexts of Fleck's thought, see Cohen and Schnelle (eds), *Cognition and Fact: Materials on Ludwik Fleck*, pp. 3–38, 161–266.
29. Nelson, 'Carl Becker Revisited: Irony and Progress in History', p. 317.
30. Becker, *Everyman His Own Historian: Essays on History and Politics*, p. 255.
31. Fleck, *Genesis*, p. 64, translation modified.
32. For discussion of the charge of self-refutation and the related charge of 'performative contradiction', see B. H. Smith, *Belief and Resistance*, pp. 73–104.
33. Becker, *Everyman*, p. 75.
34. See esp. Bloor, *Knowledge and Social Imagery*, pp. 3–23. The commitment to symmetry in the work of contemporary historians and sociologists of science – that is, to the treatment of all beliefs, true or false, scientific or non-scientific, as requiring causal explanation and as explicable in comparable ways – is the object of extensive misunderstanding, misrepresentation and condemnation by traditionalist philosophers of science and their followers. For these commentators, as for Rothstein and Lipstadt as discussed above, such efforts at evenhandedness amount to a deeply improper flattening of crucial differences and thus to an intellectually, morally and/or politically objectionable relativism. For other examples and further discussion, see Chapter 4.

35. For examples, see Chapter 1, Notes 3 and 4. The dissident philosophers include Paul Feyerabend, Nelson Goodman and Richard Rorty.
36. E. O. Wilson, for example, writes as follows:

 > The philosophical postmodernists, a rebel crew milling beneath the black flag of anarchy, challenge the very foundations of science and traditional philosophy . . . In the most extravagant version of this constructivism, there is no 'real' reality, no objective truths external to mental activity, only prevailing versions disseminated by ruling social groups . . . In the past, social scientists have embraced Marxism-Leninism. Today some promote versions of postmodern relativism that question the very idea of objective knowledge itself. (Wilson, *Consilience: The Unity of Knowledge*, pp. 40, 182)

 Of course, 'the very idea of objective knowledge itself' has been questioned since the beginning of Western thought (for example, in Plato's *Theaetetus*). For other examples of such reactions, see Gross and Levitt, *Higher Superstition*, pp. 71–106; Koertge (ed.), *A House Built on Sand*, pp. 3–6, 9–31, 59–70, 272–85; and the discussion in Chapter 5 below.
37. Proust, *Remembrance of Things Past*, vol. III, p. 953. For related discussion, see Luckhurst, *Science and Structure in Proust's* A la recherche du temps perdu.
38. Mandelbaum, *The Problem of Historical Knowledge*. The advocates of relativism specifically cited and thus answered by Mandelbaum are Wilhelm Dilthey, Benedetto Croce and Karl Mannheim.
39. For comparable responses by contemporary historians, see Himmelfarb, *On Looking into the Abyss: Untimely Thoughts on Culture and Society*; Appleby, Hunt, and Jacob, *Telling the Truth about History*; Fernández-Armesto, *Truth: A History and Guide for the Perplexed*; and, as noted above, Evans, *In Defense of History*. For comparable responses by contemporary philosophers, see Boghossian, 'What the Sokal Hoax Ought to Teach Us'; Nagel, 'The Sleep of Reason'. For comparable responses by contemporary scientists, see Chapter 5.
40. The article appeared in the *Journal of Philosophy* (cited in Novick, *That Noble Dream*, p. 165). For a contemporary counterpart, see Gross, Levitt and Lewis (eds), *The Flight from Science and Reason*.
41. Einstein's special and general theories of relativity were published in, respectively, 1905 and 1912. See Vargish and Mook, *Inside Modernism*, for plausible connections among relativity theory in physics, the Cubist movement in painting, and features such as multiple perspective, dislocated temporal sequence and indeterminacy of meaning and moral value in the novels of a number of key writers of the time. See Herbert, *Victorian Relativity*, for the recurrent effort to separate Einstein's theory from other forms of what Herbert calls 'relativity thinking', a movement that he locates in the late nineteenth century and pre-war period in England and Europe.
42. See virtually any issue of the *Journal of Philosophy* for the past thirty years. For the plague-on-both-houses claim that the objectionable relativism of the mid-century 'postpositivists' (Kuhn, Quine, Feyerabend, Bloor and others) was implicit in or anticipated by certain intellectual 'sins' of the

early twentieth-century positivists, see Laudan, *Beyond Positivism and Relativism: Theory, Method and Evidence*, pp. 3–25. As examples (of, in effect, pre-postmodern relativism), Laudan lists statements or suggestions by Popper, Reichenbach, Carnap and other Vienna Circle theorists to the effect that the methods of science could be seen as conventions; that the aims of scientists are often matters of personal taste and preference; that epistemic and other norms do not have objective validity; and that scientific theories are underdetermined by the available data. To many contemporary theorists, including a number of Laudan's fellow philosophers of science, these are all, by now, fairly well established and innocuous – or, indeed, crucially important – ideas.

43. Defensive codas or supplements and extensive replies-to-my-critics are familiar features of major works in these fields. See, for example, 'Postscript – 1969', in Kuhn, *Structure of Scientific Revolutions*, pp. 174–210; 'Afterword: Attacks on the Strong Programme', in Bloor, *Knowledge and Social Imagery*, pp. 163–86; and ' "Do You Believe in Reality?" ', in Latour, *Pandora's Hope*, pp. 1–23. For strategies of evasion, see Chapter 4 below.
44. Novick, *That Noble Dream*, p. 161. See also Herbert, *Victorian Relativity*. In economics, see, for example, Keynes, *The General Theory of Employment, Interest and Money*, pp. 16–17 and Hayek, 'Economics and Knowledge', pp. 33–54.
45. Novick, *That Noble Dream*, p. 153. See also Menand, *The Metaphysical Club*, pp. 351–75. Richard Rorty's updates and extensions of the pragmatists' views have become standard reference points for theorists in contemporary science studies and standard targets of attacks on 'postmodern relativism'. For the former, see, for example, Pickering, *The Mangle of Practice*; for the latter, see, for example, Norris, *What's Wrong with Postmodernism: Critical Theory and the Ends of Philosophy*, pp. 77–133.
46. Wimsatt, *The Verbal Icon*, p. 21. The counterpart 'Intentional Fallacy' was 'a confusion between the poem and its origins' that 'ends in biography and relativism' (ibid.).
47. See McCumber, *Time in the Ditch: American Philosophy and the McCarthy Era*. McCumber locates the shift squarely in the political anxieties of the McCarthy era, specifically as a reaction to the persecution of academic philosophers with legibly left-wing views by the government's Committee on Un-American Activities and the desertion of those thus persecuted by their departments and universities. The account is compelling but neglects other plausibly related factors, among them the intellectual interest and glamour of British philosophy for many American philosophers and the hypertrophic operation of certain values and ideals (for instance, abstractness and formal rigour) in the discipline more generally.
48. For evocations of these events and reflections on their effects on a generation of European intellectuals, see Michel Serres with Bruno Latour, *Conversations on Science, Culture, and Time*, pp. 1–42.
49. Key and/or representative works here include Barthes, *Elements of Semiology* and *Writing Degree Zero*; Culler, *Structuralist Poetics*; Foucault, *The Archaeology of Knowledge* and *Discipline and Punish: The Birth of the Prison*; Harari (ed.), *Textual Strategies: Perspectives in Post-Structuralist Criticism*; Derrida, *Of Grammatology*; White, *Metahistory: The Historical*

Imagination in Nineteenth-Century Europe; and Lyotard, *The Postmodern Condition: A Report on Knowledge*. Fredric Jameson's influential essay, 'Postmodernism, or, The Cultural Logic of Late Capitalism', first appeared in 1984.

50. Sheffler, *Science and Subjectivity*; Davidson, 'On the Very Idea of a Conceptual Scheme'; Hollis and Lukes (eds), *Rationality and Relativism*. For a spirited rejoinder to this spate of reactions, see Geertz, 'Anti-Anti-Relativism'.
51. Mohanty, *Literary Theory and the Claims of History*, pp. 10–16, 122–3.
52. The scepticism in question is illustrated by a few lines from a 1991 interview with Gayatri Chakravorty Spivak in her book, *The Postcolonial Critic*, and a passage from the 1990 collection, *Feminism/Postmodernism*, by its editor, Linda Nicholson. Citations for the recommended alternative, referred to by Mohanty as 'post-positivist realism' but more relevantly described as neo-rationalist realism, are to articles by Donald Davidson, Saul Kripke, Hilary Putnam, Charles Taylor and Richard Boyd, most of them published from the late 1970s to the mid-1980s.
53. Mohanty, *Literary Theory*, pp. 130–1.
54. Kuhn and Rorty, along with 'deconstruction' and 'poststructuralism', are indicated as crucially implicated in the relativist position in question, but no other specifically attributed instances of the position are offered.
55. Mohanty, *Literary Theory*, p. 130–1.
56. Ibid., pp. 121, 129.
57. Ibid., p. 130.
58. Ibid., p. 134.
59. Davidson, 'On the Very Idea of a Conceptual Scheme'; Taylor, 'Rationality'.
60. Davidson, 'On the Very Idea', p. 196. Amplifying the principle, Davidson writes, 'The guiding policy is to do this as far as possible, subject to considerations of simplicity, hunches about the effects of social conditioning, and of course our common-sense, or scientific, knowledge of explicable error' (ibid.), a set of considerations or loopholes that begs every question in sight.
61. For example, where Whorf writes that, in 'classify[ing] and arrang[ing] the stream of sensory experience which results in a certain world-order . . . language does in a cruder but also more versatile way the same thing that science does', and that 'all observers are not led by the same physical evidence to the same picture of the world, unless their linguistic backgrounds are similar, or can in some way be calibrated', Davidson derides a lineup of what he represents as familiar 'elements' of 'conceptual relativism', including 'language as the organizing force, not to be distinguished from science' and 'the failure of intertranslatability' (Davidson, 'On the Very Idea', p. 190 [quoting Whorf, *Language, Thought and Reality*, p. 55]).
62. See Note 52, above.
63. Winch, 'Understanding a Primitive Society', pp. 78–111. For a sophisticated update of the debate, see Latour, *Science in Action*, pp. 179–213.
64. Mohanty, *Literary Theory*, pp. 137–9, 147.
65. Ibid., p. 147.
66. See, for example, Laclau and Mouffe, *Hegemony and Socialist Strategy: Towards a Radical Democratic Politics*, pp. 93–194.

67. See Note 68 below.
68. For elaborations of poststructuralist theory in ethics, law, politics and pedagogy, including informed examinations of the conditions for effective intercultural communication and negotiation, see, for example, Spivak, *In Other Worlds: Essays in Cultural Politics* and *Outside in the Teaching Machine*; Connolly, *Identity/Difference: Democratic Negotiations of Political Paradox* and *The Ethos of Pluralization*; Brown, *States of Injury: Power and Freedom in Late Modernity*; Butler, 'Competing Universalities'; Derrida, *On Cosmopolitanism and Forgiveness*.

Chapter 3

Netting Truth: Ludwik Fleck's Constructivist Genealogy

Truth, or the diverse types of situation to which we give that name, is, for the most part, a good thing to have. It is good, certainly, when friends are loyal, lovers faithful, their tears authentic, vows earnest, stories trustworthy. It is generally in our interest to know what's up and what really happened. Not always, of course, or only: fiction and flattery, artifice and illusion, duplicity and pipe-dreams are also important, sometimes necessary, perhaps even, in their various ways, truthful, indeed sometimes supremely so – or so the poets have told us, though it's not clear they're to be trusted in such matters.

In any case, good though it is for the most part, truth seems to be in trouble these days. It is not that we are lying more or making more mistakes than in the past; the extent of those acts and ills appears pretty constant over human history. It is, rather, that certain familiar ways of thinking and talking about truth are proving troublesome. The concept appears elusive, difficult or perhaps impossible to articulate clearly in relation to other ideas – for example, *fact, reality* or *objectivity* – that have also become problematic. The term appears discursively slippery, its meanings multiple, irreducibly diverse, unstable and unfixable. Worrisome as all this is, even more troubling, from some perspectives, are assertions to the effect that this is, in fact, the case. For, it is said, such assertions (flying, as they do, in the face of truth) demonstrate the decay of intellectual competence in our time (at least in the humanities) or the domination of the academy (at least the literary academy) by dubious doctrines, such as postmodernism and radical relativism. Or, it is said, such sceptical observations about truth (even if they are to some extent true) demonstrate the failure of moral responsibility in our time. For, it is pointed out, much depends on an untroubled faith in the simplicity and stability of truth; exposing the complexities of the concept or dwelling on the ambivalent operations of the term that names it makes life easy for liars and charlatans, hard for those who know and tell the truth, and threatens to undermine the very foundations of law, education, science, history, philosophy and progressive (or, for some, conservative) politics.[1]

This situation is not, of course, altogether new. Alternatives to familiar, commonsense notions of truth have been proposed since antiquity and, also since antiquity, declared absurd, appalling or dangerous. Some of those alternatives, no doubt, were and are such. Some, however, have been and continue to be found interesting and useful, at least from some perspectives. I shall turn to these below, but, first, a few more general observations.

I

The prevailing idea of truth in both formal and informal epistemology – that is, among academic philosophers and other people who talk about such things – is that a statement or belief is true if it matches up with the way things really are, independent of anyone's statements or beliefs. This idea is quite venerable. So are certain objections to it, notably the observation that, since we cannot catch a glimpse of the way things really are around the corner of our own perceptions or descriptions, we have no way to assess statements or beliefs in regard to their reality-matching properties.[2] Various arguments have been sought and found to disarm this objection, the combined current upshot of which is that, through the dedicated pursuit of certain epistemic activities, such as rigorous reasoning, trained observation, extensive archival work, close textual analysis, or controlled experimentation, we may be brought, if not all the way to a full frontal vision of truth, then at least increasingly close to it.

The notion of truth as correspondence to the autonomously determinate features of an external reality is not only venerable but also serviceable in a wide range of informal contexts: for example, in justifying one's statements or beliefs to associates, children, untutored laypersons and perhaps oneself ('Yes, it's true, the car keys are on the hall table – go and look for yourself'). The notion does present difficulties, however, when invoked under conditions of seriously conflicting truth claims joined with seriously disparate grounds of epistemic authority and seriously divergent prior beliefs, general assumptions and relevant aims and interests. Under such conditions, the conceptual problems involved in the reality-matching theory of truth – its dubious ontological and epistemological premises and the puzzles, paradoxes and infinite regresses to which it leads when attempts are made to articulate it rigorously – manifest themselves as practical, including rhetorical, problems ('Yes, it's true, they found evidence of weapons of mass destruction – the State

Department just issued a report'). As is well known and sometimes acknowledged among professional rhetoricians (lawyers, journalists, teachers, missionaries, politicians, historians, scientists and so forth), the effective establishment of a truth, as distinct from its bare public statement, requires much hard work and many favourable social and institutional conditions, not all of which can be predicted or controlled in advance. Moreover, as is also well known among them, the task itself – that is, the establishment of a truth *as* such – can never be presumed accomplished once and for all.

II

The classic, commonsense idea of truth as correspondence to reality has been challenged since the end of the nineteenth century by a number of philosophers of generally attested intellectual competence and moral probity, including William James, John Dewey, and J. L. Austin, as well as by a number of continental (and thus perhaps naturally suspect) thinkers, such as Friedrich Nietzsche, Martin Heidegger, Michel Foucault, Jacques Derrida, and Jean-François Lyotard. Over this same period, various related, more or less highly elaborated, alternatives to that idea of truth – and, not irrelevantly here, to the entire system of assumptions, definitions and distinctions through which it operates – have been developed by scholars and theorists in fields such as the history and philosophy of science and the sociology of scientific knowledge. Some of these scholars and theorists (for example, historian/philosopher of science Thomas S. Kuhn) are/were Anglo-American. Some (such as anthropologist/sociologist Bruno Latour) are/were continental. And some are hard to place in any of these camps or categories but, perhaps for that very reason, are of special interest in connection with the current wars of truth.

Among this last group is the Polish microbiologist and medical historian Ludwik Fleck, already encountered in the last chapter. Born in Lvov in 1896, Fleck was, in his time, a distinguished scientist, specialising in immunology. His major work, *Genesis and Development of a Scientific Fact [Entstehung und Entwicklung einer wissenschaftlichen Tatsache]*, was published in Basel in 1935. As it happened, Karl Popper's important book, *The Logic of Scientific Discovery [Logik der Forschung]*, was published in Vienna the same year. Although the two works are topically related, the contrast and indeed collision of their titles could hardly be more striking; nor could the difference in their subsequent fortunes. One, Popper's, offers a *logic*: a formalisation of the supposedly general features that define genuinely scientific theories. The other, Fleck's, offers a *genealogy*: an account of the emergence of a particular medical-scientific fact and a critique of the logicism of both classic epistemology and the philosophy of science of his time. One confirms the view of science as *discovery*, the uncovering of a truth always already there. The other implies not only the historicity of truth but, no less radically, the idea that there can be a time when a scientific fact does not yet exist. The latter ideas are familiar now but were, in the 1930s, distinctly peculiar-sounding. And, indeed, though Fleck's book received some respectful, if bewildered, attention from his medical-historian contemporaries, it was Popper's *Logic* that set the terms for mainstream twentieth-century philosophy of science. I return below to the historical fortunes of Fleck's ideas, but, first, something more about those ideas themselves.

Fleck was persuaded by his clinical and laboratory experiences as well as by his readings in medical and intellectual history that scientific facts are not prior, fixed and autonomously determinate features of an external world but, rather, as he puts it, 'event[s] in the history of thought'.[3] As he saw it, the emergence and specific features of such events – what we come to speak of as facts – are made possible but also severely constrained by the social-psychological operations of particular 'thought styles' [*Denkstilen*]: that is, systems of ideas and assumptions and related perceptual, classificatory and behavioural dispositions that prevail among the members of particular epistemic communities (scientific fields, academic disciplines, religious sects and so forth) or, in his term, 'thought collectives' [*Denkkollektiven*]. 'Truth,' he writes,

> is not 'relative' and certainly not 'subjective' in the popular sense of the word. It is always, or almost always, completely determined within a thought style. One can never say that the same thought is true for A and false for B. If A and B belong to the same thought collective, the thought will be either true or false for both. But if they belong to different thought collectives, it will just *not* be *the same* thought! It must either be unclear to, or be understood differently by, one of them. Truth is not a convention, *but rather (1) in historical perspective, an event in the history of thought, [and] (2) in its contemporary context, stylized thought constraint.*[4]

As Fleck goes on to explain, there are comparable constraints in every field of human production: for example, in music, literature and painting, where not all stylistic choices are available at a given time to a composer, writer or artist. In science and other fields of knowledge-production, the operation of such constraints is experienced by members of the collective as 'a *signal of resistance* to unconstrained, free, arbitrary thinking' and interpreted by them as reality, self-evident fact, or objective truth.[5] To be even perceptible, however, a fact must be in harmony with the prevailing thought style and aligned with the intellectual interests and other goals – for example, technological projects – of the relevant community. These two sets of constraints – limitation of choices by communal style and prevailing interests – are experienced and interpreted by members of the community as 'objective connection[s] between phenomena . . . conditioned only by logic and content' and otherwise unmediated.[6] Because historians of science and writers of science textbooks participate in this post-hoc experience and interpretation, standard descriptions of the 'discovery' of scientific facts obscure the social, institutional and cognitive processes involved and make it appear as if those contingent connections existed a priori or were given in the nature of things. The only way to make those processes visible, Fleck believed, would be by means of a new field of study – 'comparative epistemology' – that traced the historical emergence and development of facts in their intellectual and social contexts.

The centrepiece of Fleck's book is a richly detailed, often hands-on account of the genesis and development of a specific complex fact, the so-called Wassermann Reaction, that is, an observable change in a chemically treated blood sample that indicates the presence of the syphilis pathogen. As Fleck recounts the story, the otherwise invisible processes and connecting lines that give this fact its scientific solidity come into sharp and often surprising focus. Specifically, one sees how the individuating features of each of the three components of the Wassermann Reaction (that is, the disease entity, the blood-testing procedure and the pathogenic micro-organism, *Spirochaeta pallida*) emerge from the *reciprocally shaping and sustaining* activities – observations, hunches and experimental manipulations – of countless physicians, chemists, biologists and laboratory technicians, as shaped by shifting popular and specialised beliefs about sex, sin, punishment, affliction and the workings of the human body, coupled with changing tools and techniques of medical diagnosis and treatment, coupled with changing theories and related methods in chemistry and bacteriology, coupled with differently motivated but convergent public, political and professional interests in diagnosing a certain ailment. Fleck draws the moral of his story this way:

> Historically, [the Wassermann Reaction] appears as the only possible junction of the various trains of thought. The old idea about the [tainted] blood [of syphilitics] and the new idea of complement fixation [as a diagnostic method] merge in a convergent development with chemical ideas and with the habits [of perception and behaviour] they induce to create a fixed point. This in turn is the starting point for new lines everywhere developing and again joining up with others. Nor do the old lines remain unchanged. New junctions are

> produced time and again and old ones displace each other. This network in continuous fluctuation is called reality or truth.[7]

A network in continuous fluctuation – a rather striking metaphor for truth or reality! It may be compared with the epigraph of Popper's *Logic of Scientific Discovery*, a line from Novalis: 'Theories are nets: only he who casts will catch.'[8] For Popper, the net, an individually conceived conjecture, may *catch* truth. For Fleck, the net, a web of shifting, intersecting, interacting beliefs and practices, *is* truth.

According to Fleck's central idea here, the statement or belief that we call truth and may experience and describe as corresponding to reality might be better described as extensively and effectively linked to and congruent with what we otherwise experience as stable, resistant and real. Thus, another of his metaphors for systems of interrelated beliefs, perceptions and actions – this one a bit mischievous – was 'harmony of illusions'.[9] The mutual shaping and coordination of perceptual, conceptual and behavioural practices; a stable and effective congruence among ideas, observations and manipulations; a consonance among beliefs, perceptions and actions: none of these comes down to the matching of beliefs or statements to an independent external reality – that is, to a classic or commonsense idea of truth. But neither does any of them come down to a *denial* of external reality in the sense of something other than our own statements, beliefs or experiences with which we interact. The central ontological/epistemological implication of Fleck's work and of constructivist thought more generally is not that there is nothing 'out there'. It is, rather, that the specific features of what we interact with *as* reality are not prior to and independent of those interactions but emerge and acquire their specificity *through* them.

In a compelling passage early in the book discussing what he calls 'proto-ideas' (for example, ancient Greek notions of elements and atoms, the medieval conception of syphilis as foul blood, or early rough ideas of micro-organisms), Fleck rejects the image of science as winnowing 'true' ideas from 'false' ones:[10]

> Implicit in such a view is the dubious claim that the categories of truth and falsehood may be applied to these proto-ideas . . . [But] their accuracy, truth, and value . . . cannot possibly be determined outside their particular contexts . . . as they [these proto-ideas] were produced within a thought collective, different from our own, in a thought style different from our own . . . A general criterion of correctness for fossil theories is no more appropriate than an absolute criterion of adaptability for paleontological species. The brontosaurus was as suitably organised for its environment as the modern lizard is for its own. If considered outside its proper environmental context, however, it could not be called either 'adapted' or 'unadapted'.[11]

For Fleck, it is not a matter of wrong ideas or false theories being corrected, disproved or superseded by science, itself conceived as a monolithic epistemic agent, process or storehouse. Rather, ideas that are more or less workable but still vague become refined, transformed and connected to other more or less workable ideas and, in this way, develop over time into what are accepted as scientific theories – but which, as such, are subject to further transformation. Just as there can be no absolute criterion of fitness for biological species, so there can be no general criterion of truth for ideas, no way of assessing their epistemic value independent of the intellectual environments in which they emerged and to which they were 'adapted'.[12]

III

Genesis and Development is actually two genealogies. One is a cultural and intellectual history of the development of the concept of syphilis from the medieval European notion of a 'carnal scourge' to its early twentieth-century identification as a specific disease entity associated with a specific microbial pathogen. The other is a detailed account of the establishment of the Wassermann Reaction – that is, the diagnostically useful correlation between certain observable changes in a chemically treated blood sample and the presence of the syphilis pathogen. Fleck's systematic reflections on the epistemological implications of these two interrelated histories amount to a highly original theory of cognition – individual, collective and scientific – and, as such, operate as a strong and far-reaching challenge to classic views of knowledge. Moreover, as we can see more clearly now than was evident to his contemporaries, but as was certainly sensed by Thomas S. Kuhn when he came upon Fleck's book in the late 1940s, they offer a compelling set of alternatives to both conventional intellectual history and conventional philosophy of science.

Genesis and Development was a remarkably innovative work in its time and remains one of the most theoretically radical texts in twentieth-century epistemology and philosophy of science. Although not itself well known, its intellectual impact has been substantial, largely as mediated by Kuhn's *Structure of Scientific Revolutions*, which adopted many of Fleck's most original and challenging ideas and, since its own publication in 1962, made them familiar to a large segment of the academic community and beyond. Kuhn's 'paradigms' and 'scientific communities', for example, bear a more than passing or coincidental resemblance to Fleck's 'thought styles' and 'thought collectives'.[13] By way of Kuhn joined with Wittgenstein and also more directly, *Genesis and Development* was a major influence on the Strong Programme in the sociology of knowledge and related work in the history of science.[14] By way of Kuhn joined with poststructuralist theory and, again, also more directly, Fleck's work has been important for corresponding and often convergent developments on the continent, including the 'actor-network theory' of techno-science associated with Bruno Latour and Michel Callon.[15] Indeed, Latour's *Pasteurization of France*, which recounts the formation, triumph and dissemination of the microbe theory of disease, could be read as an epic – or mock-epic – extension of Fleck's narrative of the Wassermann Reaction.

Fleck's reconceptualisations of *truth*, *facts*, *science* and *reality* are not simple. Nor can the gist of his thought be given in a phrase or two. The articulation and elaboration of these ideas in *Genesis and Development* involve a number of novel concepts and conceptual connections, a good bit of substantively significant historical, sociological and technical detail, the working through of a series of important theoretical implications and the framing of replies to a range of potential challenges and questions. If one turns the pages of Fleck's book searching for assurances of the reality of Reality or the ultimately objective nature of scientific truth, one will not find them. If one searches it through for possibly scandalising statements, one will find a good number of them, such as those already quoted here. If one's engagement with Fleck's text and with the work of the historians and sociologists influenced by his thought is confined to searches of that kind, then one's understanding of constructivist views of knowledge and science will be quite limited. (One shall, however, be equipped to speak authoritatively to similarly uninformed audiences regarding the absurdities and perils of 'postmodern' science studies, including its 'attacks' on science, its 'denial' of reality and its disdain for 'the very idea of truth'.[16]) On the other hand, there may be no better way to begin to appreciate the power of constructivist views of knowledge and their role in contemporary science studies than through an attentive reading of *Genesis and Development*. Such a reading, as I hope to indicate here, offers a range of other rewards as well – though not commonly, in the end, the pleasures of being scandalised.

IV

'Must we conclude that epistemology is not a science?'

As depicted by August von Wassermann, the discovery of the reaction that bears his name was the result of a set of increasingly successful experiments by an individual scientist, with a specific goal, engaged in a highly focused search. To Fleck, however, this image of scientific agency and alignment of intentions, actions and outcomes is clearly the product of retrospective selection and schematisation. If one looks at the record of the events leading to the discovery, one sees not a straight line but 'a meandering path' that includes false assumptions, vague hunches, unsuccessful experiments, lucky accidents and, at every point, the contribution of useful ideas, methods and technical adjustments by many different people.[17] Moreover, nothing about this is unusual; the discovery of the Wassermann Reaction is a paradigm of how scientific discovery occurs.[18]

Paul Feyerabend makes a similar set of points thirty years later in his epistemologically 'anarchist' treatise, *Against Method*. When we look at the historical record of the actual activities of scientists making discoveries, he observes, including such models of scientific success and propriety as Galileo's discovery of the moons of Jupiter, we find, first, that there is no particular method that characterises those activities (on the contrary, they are multiple, various and opportunistic) and, second, that they always include much that would be considered extra-scientific or downright unscientific (for example, social, political and rhetorical efforts). If we require a methodological rule for producing scientific knowledge, Feyerabend concludes provocatively, we had better make it 'Anything goes!'[19] Fleck's observations lead to a similarly provocative question. If, as it appears, truth is routinely discovered by error and accident, then how, he asks, can epistemology, supposedly the study of the grounds of valid knowledge, proceed? Must we conclude that epistemology is not a science? '*Ist denn Erkenntnistheorie keine Wissenschaft?*'[20]

As articulated most influentially by the theorists of the Vienna Circle, the proper answer to Fleck's question is: *No*, epistemology (*Erkenntnistheorie*) is not a science; it is a branch of philosophy. As such, however, it can establish criteria for distinguishing science from non-science, determine the logical propriety of claims to scientifically valid knowledge, and advise scientists (and other people) on the best ways to go about discovering truth. Fleck's answer to the question is different on every crucial point: *Yes*, the theory of knowledge (*Erkenntnistheorie*) is, at least potentially, a scientific enterprise – but, to be pursued as such, its methods must be empirical, not exclusively logical; its goal must be to explain the phenomena of cognition adequately, not to tell good from bad science or right from wrong reasoning; and the successes of science must be examined as *social*, not individual, achievements.[21]

The problem that concerned Fleck most centrally was not the classic 'What is knowledge?' or 'How do we certify that we know something?' but 'How does that which we call knowledge come into being?'

Exploring that problem meant seeking to understand the mechanisms of cognition at every level: those of the individual subject (scientist or layman) in the course of his or her lifetime; those of the social collective (scientific discipline or other field of knowledge) over several generations; and those that characterise science per se, as a specific sort of technical-cognitive enterprise, over historical time. Accordingly, Fleck conceived the new field of comparative epistemology as an exceptionally comprehensive enterprise. Multi-levelled and interdisciplinary, it included experimental studies in psychology, research in anthropology and sociology, archival work in cultural, social and intellectual history, and observations of science and scientists onsite and in action. In his descriptions of the project and illustrations of it in *Genesis and Development*, Fleck drew no line between philosophy and the social sciences, or between psychology and sociology, or between any of these and social, political or intellectual history. Each of these divisions, however, became institutionally significant in the decades that followed, and the various disciplines involved became increasingly mutually isolated and to some extent antagonistic. Thus, although a number of Fleck's central observations, such as the rigidity of belief systems or the effect of prior expectations on perception, would be studied experimentally by social and cognitive psychologists, the two fields most immediately and extensively influenced by his work, the history and sociology of science, developed along lines quite different from – and often in determined contradistinction to – those of empirical psychology. Moreover, while the history and sociology of science, especially as assembled under the label 'science studies', became increasingly explicitly constructivist,[22] cognitive psychology, especially joined with artificial intelligence under the label 'cognitive science', maintained close connections to rationalist philosophy of mind and, in some places, operated as a stronghold of traditional epistemological assumptions and ambitions.[23] For these reasons and related ones discussed later in this chapter, Fleck's comparative epistemology had little chance of uptake through most of the twentieth century. As cognitive science, social studies of science and philosophy of science become increasingly interconnected in our own era, however, something like that project may yet be realised.[24]

V

The notion of truth, facts or reality as the product of a 'harmony of illusions' is one of Fleck's most important ideas, but also perhaps the most scandalous one, especially in the context of classic dualisms of truth and error, reality and appearance, and correspondence and disparity. For Fleck, pursuing a thoroughly naturalised epistemology, truth and facts are emergent effects, the products of general psychological tendencies and culturally and historically contingent social processes.[25] The questions that interest him are how it comes about that something – object, entity or state of affairs – appears what we call 'real' or that certain statements seem well supported or self-evident while others seem clearly wrong or absurd. Both in the questions he asks and the answers he frames, 'appearance' and 'seeming' are not opposed to 'reality' or 'being', but neither are they identified with each other: all these are seen, rather, as interpretations of experiences that are *more* or *less* collectively attuned, *more* or *less* stable and *more* or *less* pragmatically reliable.

As Fleck explains the phenomenon, the harmony of illusions is a product of powerful cognitive tendencies, among them what he calls 'the tendency to inertia' of belief systems. 'Once a structurally complete and closed system of beliefs [*Meinungssystem*] consisting of many details and relations has been formed, it offers tenacious resistance to anything that contradicts it.'[26] This tendency is not a pathology of thought, but the way all thought operates – in effect, a cognitive universal, but one with ambivalent features.[27] On the one hand, 'when a conception permeates a thought collective strongly enough, so that it penetrates as far as everyday life and idiom and has become a viewpoint in a literal sense of the word, any contradiction appears unthinkable and unimaginable.'[28] For example:

> The prevailing associations of childhood with purity and of sexuality with adulteration have rendered children's sexuality imperceptible . . . [in spite of] the fact that everyone has the experience of being a child and later lives not altogether isolated from them.[29]

On the other hand, 'only a classical theory with associated ideas which are plausible because rooted in a given era . . . and communicable because stylistically relevant, has the strength to advance.'[30] In established belief systems, coherence among perceptions, beliefs, background assumptions and material practices are preserved by *ongoing mutual adjustment*. The sense (impression, conviction) of the truth or validity of some statement can be understood, accordingly, as the experience of its consonance – harmony – with other perceptions, accepted statements, general assumptions and embodied recollections of past and/or current interactions. This experience of harmony is an illusion insofar as we project it outward and regard it as an objective correspondence of statement and world, independent of the cognitive processes from which it emerged and of the other elements that sustain its coherence. The experience is also an illusion insofar as we explain it as the product of some privileged cognitive process (such as logic or reason as classically conceived) or some set of putatively orthotropic procedures (such as a particular 'scientific' method). But it is not an illusion in the sense that there is some set of otherwise verifiable experiences that contradicts it or some otherwise cognisable reality to which it fails to conform.

Fleck goes on to discuss what he refers to as the most active or 'creative' aspect of this inertial tendency of belief systems, 'the so-to-speak magical realization of ideas, the explanation of how the dreams of science are fulfilled'.[31] The history of science, he notes, is filled with stories about how a certain imaginative speculation or prediction was verified in fact – for example, the existence of a branch of the uterus corresponding to the ejaculatory duct, as predicted by Vesalius.[32] But, Fleck continues, this is because observations of the data were unwittingly shaped to conform to the style of the relevant belief system. No such branch of the uterus as depicted in the old anatomy books is known in modern anatomy. Moreover, contemporary depictions of the uterus are no less stylised than the old ones, though in a different way.[33] His commentary here is still powerful:

> It is true that modern doctrine is supported by much more sophisticated techniques of investigation, much broader experience, and more thorough theory. The naïve analogy between the organs of both sexes has disappeared, and far more details are at our disposal. But the path from dissection to formulated theory [and pictorial representation] is extremely complicated, indirect, and culturally conditioned . . . *In science, just as in art and in life, only that which is true to culture is true to nature.*[34]

VI

Fleck was a mordant critic of the often crude empiricism of the epistemological tradition, especially as perpetuated in the philosophy of science of his era. Standard descriptions of scientific observation, he writes, are subject to a popular myth about how cognition operates:

> The knowing subject acts as a kind of conqueror, like Julius Caesar winning his battles by the formula *veni-vidi-vici*, 'I came, I saw, I conquered'. One wants to know something, makes the observation or does the experiment – and already one knows it . . . But the situation is not so simple . . . [O]ne cannot observe or ask questions properly . . . until tradition, education, and familiarity have produced *a readiness for stylized (that is, directed and restricted) perception and action.*[35]

Fleck notes that Rudolph Carnap's *The Logical Structure of the World* (1928) was perhaps 'the last serious attempt to construct the "universe" from "given" features and from "direct experience" construed as the ultimate elements'.[36] Criticism of the idea is unnecessary, he continues, since Carnap himself has already relinquished it. He adds:

> [O]ne would hope that eventually [Carnap] might discover the social conditioning of thought. This would free him from absolutism regarding standards of thought, but of course he would also have to renounce the concept of 'unified science'.[37]

The discovery and liberation (and related renunciation) evoked here did not take place, of course, at least not among Vienna Circle theorists or in the philosophical tradition they defined. Logical positivist/empiricist assumptions, 'absolutist' normative ambitions and convictions of the ultimate unification of science continued to dominate mainstream epistemology for the next fifty years, along with the belief that 'the social conditioning of thought' was an obstacle to truth, not, as Fleck argued, a key to understanding it.

The old epistemological dream was that, by some combination of rigorously rational methods – radical scepticism, reducing observation statements to incontestably self-evident elements, analysing, factoring out, bracketing and so forth – one would be able to eliminate (or at least minimise) the distorting, obscuring effects of sensory error, personal bias or social influence and arrive at (or at least approach) certainty in knowledge. In the philosophy of science, this became the search for a set of orthotropic methods – logically proper formulation and empirical testing of hypotheses, trained observation and controlled experimentation, the use of objective instruments of measurement and recording and so forth – to the same end: to cancel out the personal, subjective, social and political; to arrive at an accurate, undistorted representation of nature; to net truth itself. Fleck was not drawn to that dream. On the contrary, he was persuaded that scientific observation is inevitably shaped by collective assumptions and ongoing social coordinations and, furthermore, that such shaping and coordination are inextricable elements of the process that yields scientific facts.

VII

The term 'thought styles' is somewhat awkward and perhaps naïve sounding in English. Related but more familiar and perhaps more sophisticated sounding notions include 'discourses', 'regimes of truth', 'language games' and, of course, 'paradigms'. Although they are not equivalent, each of these terms points to the existence and significance of established systems of linked assumptions, convictions, values and discursive-technical practices – and, for Fleck, related perceptual and cognitive dispositions. There is no such thing, he insists, as *unstylised* – 'direct', 'pure', 'objective' – sensation, perception, conceptualisation, description or knowledge. A thought style is a disposition not merely to think or speak but also to perceive one way rather than another. Thus, for the members of a collective who share a given thought style, certain entities, categories, and connections will be especially salient and ready-to-hand and others less noticeable or invisible. These perceptual-conceptual dispositions are not 'biases', a term that suggests disabling distortions of otherwise clear or direct perceptions. Rather, and precisely because of how they *constrain* cognition, such dispositions *enable* what we call facts to be known, what we call reality to be brought forth and experienced.

Responding to the charge that the idea of a group mind (implied by the notion of a collective thought style) is 'metaphysical', Fleck replies that, yes, the idea is a hypostatised fiction – but what concept, scientific or otherwise, isn't? Is the idea of an individual person, he asks, or an individual mind – or, for that matter, the structure or form of an individual 'body' – any less the reification of a set of functions?

> How does one arrive at the structure [*Gebilde*] of 'body' as a specific form to be directly perceived? There is no doubt that in everyday life, with the several senses of feeling, pain, muscles, vision, we actually perceive 'bodies' without any difficulty . . . But upon analysis these 'bodies' dissolve into functions.[38]

He adds a strong ontological (or anti-ontological) reflection:

> The boundary line between that which is thought and that which is [taken to exist] is too narrowly drawn. Thinking must be accorded a certain power to create objects, and objects must be construed as originating in thinking; but, of course, only if it is the style-permeated thinking of a collective.[39]

Thought styles are crucial to scientific observation and discovery: the condition, it could be said, of their possibility. In the absence of the habits of style-directed perception gained by disciplinary training and experience, a researcher's observations in the conduct of an investigation are vague and uncertain:

> Confused partial themes in various styles are chaotically thrown together. Contradictory moods have a random influence . . . Nothing is factual or fixed. Things can be seen almost arbitrarily in this light or that. There is neither support, nor constraint, nor resistance, and there is no 'firm ground of facts'.[40]

By the same token, however, the effects of thought styles on perception are the condition of *impossibility* of certain observations and discoveries. Thus, during the classical period in bacteriology (when the views and procedures of Koch and Pasteur were dominant), 'the all-pervasive power of practical success and personalities created a rigid thought style.' '[O]nly a strictly orthodox method was recognized and the findings were correspondingly very restricted and uniform.'[41] Because researchers during that period regarded all secondary changes in bacteria cultures as 'pathological' or 'artificial' phenomena, they ignored what bacteriologists now recognise as the fact of morphological variability within species. 'If a [bacteriologist] of that time had been asked why this principle was accepted or why the characteristics of species were conceived this way, he could only have answered, "because it is true".'[42] The fact could no longer be ignored, however, after detailed observations of the variability of a particular species were made under controlled conditions and, no less significantly, 'couched throughout in terms of the current thought style'.[43] Fleck's point here, once again, is that there are no pure observations, complete descriptions, or 'raw' data. All observations, including *or especially* those of a highly trained, extensively experienced scientist, are shaped and selected by prior belief and experience – that is, by the ideas, assumptions and practical know-how that, operating together, induce the perceptual expectations and perceptual-behavioural dispositions that, duly mutually adjusted among the members of a collective, yield what we call (scientific) knowledge.

Elaborating these points, Fleck introduces a vivid metaphor from musical improvisation. Noting that, in Wassermann's laboratory, the indicators in the early experiments with syphilitic blood samples were vague and ambiguous, he writes:

> [But it is clear] that in these muddled notes Wassermann heard the melody that hummed within him but was inaudible to those not involved. He and his co-workers listened and tuned their instruments to the point where these [notes] became selective and eventually the melody could be heard even by ordinary laymen.[44]

Like the uncertain and scattered opening riffs of an improvisatory jazz group, the early experiments performed by Wassermann and his associates initiated a process of *selective production* and *mutual attunement*.[45] Here and elsewhere, the discovery – or construction or emergence or development – of a scientific fact is a process of gradual co-adjustment of assumptions, expectations, procedures and observations among a group of people with common aims, a shared tradition, a history of collaboration, and (in this, too, like the members of a jazz group) a shared perceptual-cognitive-performative style. Nothing in this process could be called 'true' or 'right' per se. But the product of this process, a densely woven network of extensively interconnected and mutually supportive elements, is a harmonious, satisfying, effective structure – a conceptually coherent, perceptually stable, pragmatically reliable set of ideas and practices experienced as right and fitting by the members of a community.[46]

VIII

Thomas Kuhn, distancing himself from Fleck's thought in his foreword to the 1979 edition of *Genesis and Development*, refers to the 'vaguely repulsive perspective of a sociology of the collective mind'.[47] Kuhn was probably alluding here to the association of socially conditioned thought with distortion or inculcation and perhaps to various studies, both before and after the Second World War, of 'mob' or 'mass' psychology.[48] Such associations had been given substance in the 1930s and 1940s by images of crowds chanting and saluting in unison, wildly cheering brutal leaders and racist, jingoist speeches – images that would have been especially vivid for, among others, the émigré generation of Austrian and German philosophers of science and, significantly here, would have reinforced their convictions of and efforts to demonstrate a fundamental difference between science and ideology, truth and propaganda, reason and irrationality.

Writing in the early 1930s, Fleck is aware of such associations but rejects the related convictions, distinctions and efforts. He cites Gustave Le Bon's account, in his *The Crowd: A Study of the Popular Mind*, of a sighting by the entire crew of a ship of a small boat in distress with passengers calling and waving their arms for help, later identified by the crew as a drifting tree.[49] Commenting on the story, he stresses the similarities of such effects to scientific discovery – as well as some important, but not conventionally framed, differences between them:

> This case could be considered the very paradigm of many discoveries. The mood-conforming gestalt-seeing and its sudden reversal: the different gestalt-seeing . . . The same situation obtains in scientific discovery, only translated from excitement and feverish activity to equanimity and permanence. The disciplined and even-tempered mood, persisting through many generations of a collective, produces the 'real image' in exactly the same way as the feverish mood produces a hallucination. In both cases a switch of mood (or switch of thought style) and switch of image proceeded in parallel.[50]

Fleck's reference to the reversal of perceptions here as 'a different gestalt-seeing' (rather than, say, as the crew's discovery of the truth) prefigures and perhaps influenced Kuhn's later and ultimately notorious invocation of 'gestalt-switching' in visual illusions to help characterise paradigm shifts in scientific revolutions.[51] The analogy aroused the horror of some mid-century American philosophers of science, who accused Kuhn of suggesting that scientists' choice of one theory over another was 'irrational'.[52] Evidently disturbed by the charge, Kuhn insisted that he did not intend such a suggestion and, in the course of restating his view, weakened some of its most challenging features.[53] Fleck would presumably have responded differently, at least to the immediate charge; for his point here, as throughout *Genesis and Development*, is that an adequate understanding of cognition makes classic conceptions of both reason and irrationality obsolete.

Fleck sees highly contingent cultural and political factors as significant for the emergence of scientific facts, but again, as with social 'conditioning', not as sources of bias or distortion. The Wassermann Reaction, he insists, is certainly a scientific fact; that is, there is certainly a diagnostically useful correlation between a particular change in the appearance of a treated blood sample and the presence of the syphilis pathogen. But that fact would never have been discovered if the European public had not been so anxious about syphilis; and they would not have been so anxious about the disease were it not for its ancient religious-metaphoric-emotional status as 'the carnal scourge' and its popular association with the sex act, heredity and moral degeneration. (Fleck notes that tuberculosis, which claimed many more victims at the time, did not receive nearly as much attention.) Nor would the reaction have been discovered if Wassermann, working in a state-supported research institute in Germany, had not been pressured by a minister of health who, aware of recent advances in experimental biology in France, was conscious of national honour. These historically conditioned cultural anxieties and political concerns operated together to make the search for a diagnostic test for syphilis a significant scientific project in Germany at the beginning of the twentieth century. They also operated to provide a high degree of public energy, popular interest and institutional support for its intensive pursuit. Nothing in this description of the significance of 'external', 'non-epistemic' forces suggests that the status of the Wassermann Reaction as a scientific fact was thereby compromised. To acknowledge, trace and specify the inevitable and indeed crucial operation of cultural, social and political forces in the development of scientific knowledge is not to claim (or charge or concede) that science is fundamentally biased, corrupt or in the service of power – though, of course, the effects of such forces are not always intellectually or socially benign either.[54]

IX

> [C]ognition must not be construed as only a dual relation between the knowing subject and the object to be known. The existing fund of knowledge must be a third partner in this relation as a basic factor of all new knowledge.[55]

Fleck's elaboration of this rejection of the subject-object relation in cognition amounts to an epistemic relativism as flagrant today as it was in 1935. '[J]ust as the statement . . . "Town A is situated to the left of town B" is incomplete and demands an addition such as "to someone standing on the road between towns A and B facing north," ' so, analogously, he writes,

> the statement, 'Someone recognizes something', demands some such supplement as 'on the basis of a particular fund of knowledge', or, better, 'as a member of a particular cultural environment', and, best, 'in a particular thought style, in a particular thought collective'.[56]

For example, he continues, the statement '[Schaudinn] discovered *Spirochaeta pallida* as the causative agent of syphilis' is equivocal or meaningless without its particular historical-intellectual context. 'Torn from this context,' Fleck insists, ' "syphilis" has no meaning and "discovered" by itself is no more explicit than . . . "[to the] left" in the examples above.'[57]

As it happened, another scientist, Siegel, also discovered protozoa-like structures and suggested that they were the causative agent of syphilis. 'If his findings had had the appropriate influence and received a proper measure of publicity throughout the thought collective,' Fleck writes, 'the concept of syphilis would be a very different one today.'[58] A different set of diseases would be classified together as syphilis and a different idea of infectious disease would have arisen. 'Ultimately we would also have reached a harmonious system of knowledge along this line, but it would differ radically from the current one.'[59] Having laid out this hypothetical alternative in some detail, Fleck goes on to argue that, though logically possible, it is a 'historical impossibility':

> At the time Siegel made his finding, the concept of syphilis was too solidly established for such a sweeping change to occur, and a hundred years earlier, when the concept was still sufficiently adaptable and fluid, the intellectual and experimental-technological conditions necessary for such a finding did not yet exist. We need have no scruples about declaring Schaudinn's finding correct and Siegel's incorrect: for the former was uniquely (or almost uniquely) connectable with the thought collective, whereas the latter lacked such a connection.[60]

Two major points are being made here. One is that the acceptance of a theory as correct is not accidental, arbitrary or just the product of social convention,[61] but depends on the extent to which it can be linked to ideas already established in the relevant community within a more general intellectual and cultural, including technological, context. This is very different, of course, from saying that a theory's acceptance is explained by its rational assessment or by its duly demonstrated conformity to the evidence – that is, by what are commonly seen as the scientifically legitimate and *only* alternatives to 'arbitrary choice', 'historical accident' or 'mere social or political interests'. The second point, however, is that, *for that very reason*, it makes no sense to entertain historically counterfactual scenarios in which some rival theory was accepted as correct or, beyond that, to speak of alternative theories as equally valid.[62] Thus two claims usually identified or seen as logically entailed are here seen as quite distinct, with one *prohibited* by the other: *because* (1) the validity of a theory depends on its position in a network of historically specific connections, *it cannot be the case that* (2) alternative theories are equally valid.

Bruno Latour makes a similar set of points seventy-five years later in *We Have Never Been Modern*, though more obliquely. Latour begins by distinguishing between an objectionable '*absolute* relativism' that denies the existence of any common or transcendental yardsticks and a commendable '*relative* relativism' that, taking its name seriously, recognises the central significance of *relations* and 'rediscovers . . . the process of establishing' them:

> To establish relations; to render them commensurable; to regulate measuring instruments; to institute metrological chains; to draw up dictionaries of correspondences; to discuss the compatibility of norms and standards; to extend calibrated networks; to set up and negotiate valorimeters – these are some of the meanings of the word 'relativism'.[63]

He draws this moral:

> A little relativism distances us from the universal; a lot brings us back, but it is a universal in networks . . . [T]he universalists defined a single hierarchy. The absolute relativists made all hierarchies equal. The relative relativists, more modest but more empirical, point out what instruments and what chains serve to create asymmetries and equalities, hierarchies and differences.[64]

In Latour's account as in Fleck's, the acceptability of a scientific theory depends on the strength, extent and stability of its linkages in an established cognitive-technological network: its being connectable to already accepted ideas; its congruence with already accepted findings; its measurement by available instruments and techniques; its description in the prevailing conceptual idiom; its recurrent communication through established channels of social interaction; and its effective application in ongoing projects and practices. In both accounts, this means that the operative validity of a scientific theory does not exist – will not be experienced as such – outside such a network and cannot be described or assessed without implicit reference to one. In neither account, however, does it follow that all theories are equally valid. On the contrary, what does follow is that, in relation to a given network at a given time, linkages of the sort just mentioned (connections, congruences, communications, applications and so forth) will be experienced by members of the relevant communities as stronger, more stable and more extensive for one theory than for another. Thus, as Latour insists, the epistemic 'symmetry' or presumptive 'equal credibility' of theories posited by the absolute relativist is strictly hypothetical. It is always broken in fact – made asymmetrical and hierarchical – by the particulars of history.[65]

X

Unlike Kuhn and to some extent Foucault, Fleck does not invoke radical discontinuities – 'revolutions' or 'ruptures' – in his accounts of intellectual history. To be sure, like each of them, he rejects traditional conceptions of scientific knowledge as cumulatively progressive and, like them, stresses both the contingency of the emergence of individual disciplines (or 'discourses') and the disparity and possible incommensurability of concepts and conceptual styles (or 'epistemes' or 'paradigms') in different eras. Fleck sees all these, however, as matters of ongoing more or less extensive and fundamental transformation across multiple dimensions, in multiple domains, rather than, as in Kuhn's *Structure*, of individually discrete cycles of normality, crisis and revolution. This does not mean that Fleck's views of intellectual history are less theoretically radical than Kuhn's. What it indicates, rather, is that they are attentive to subtler and more heterogeneous processes and effects. In this respect as in others, Fleck's thought is closer to Foucault's than to Kuhn's.

These points can be amplified a bit. The idea of discontinuities in intellectual history are emphasised by Foucault in his earlier works (notably *Madness and Civilization*, *Birth of the Clinic*, and *The Order of Things: An Archaeology of the Human Sciences*) as part of a wide-ranging critique of conventional intellectual history, including its assumptions of smooth progress and what he calls 'the teleology of reason'. By the time he was writing *The Archaeology of Knowledge* (1969), however, with its focus on the multiple, heterogeneous strata and variable tempos of intellectual change,[66] he was obviously finding ideas of 'rupture' too simple and such emphasis no longer as important. Although it is doubtful that Foucault knew *Genesis and Development* directly, there are extensive affinities between his thought and Fleck's. Like Fleck, Foucault rejects the rationalism of classic epistemology and reconceptualises the 'facts' and 'objects' of institutionalised scientific knowledge as reifications naturalised through powerful and pervasive social processes. Thus Foucault's accounts of the social construction of such 'discursive objects' as 'criminality', 'imbecility' and 'sexual perversion' in nineteenth-century psychiatry parallel quite closely Fleck's accounts of the emergence of the particular disease-entity 'syphilis' and of the very idea of 'disease' in Western medicine. Along with Latour's account of the emergence of such 'quasi-objects' as 'microbes' in Pasteur's laboratories,[67] these can be seen as paradigmatic examples of contemporary constructivist (anti-) epistemology/(anti-)ontology.

There are, however, also major differences in their work. Fleck's laboratory-instructed accounts are more attentive than Foucault's to the significance of technical practices, such as measurements, manipulations or the skills and habits of laboratory technicians. Also, with perhaps a stronger interest in psychological phenomena than Foucault or a less sceptical conception of the field of psychology,[68] Fleck was more attentive to the operation of specifically cognitive-perceptual forces in the social construction of scientific knowledge. On the other hand, Foucault's distinctly poststructuralist accounts are considerably more sensitive to matters of language and to the significance of discursive practices (description, enumeration, classification, codification, labelling and so forth) in the construction not only of 'objects', such as 'imbecility' and 'sexual perversion', but also of *subjects*, such as 'imbeciles' and 'perverts'. Most significantly, perhaps, while Fleck sought centrally to expose the disparities between prevailing ideas about scientific knowledge and what he observed as the processes of its production, Foucault's concerns extended to the often dubious epistemic authority of science itself, especially that of 'the human sciences' (medicine, psychiatry and so forth), and to the collaborations of the latter with political authority and established social interests. Experience suggests that these differences are liable to invidious reframing – as, for example, Fleck's concern with '(real, material) *things*' versus Foucault's with '(mere, insubstantial) *words*' or Fleck's '(merely/genuinely) *intellectual* interests' versus Foucault's '(genuinely/merely) *political*' ones. But nothing of either intellectual or political value is gained, I think, by such characterisations and assessments, which reinstate the conventional dualisms and hierarchies that both Fleck and Foucault devoted themselves to disrupting and obliterate the alternative conceptualisations that each of them laboured to achieve.

XI

As Fleck describes them, thought collectives are communities of interacting individuals who share systems of beliefs and practices: for example, astrologers, Lutherans, eighteenth-century European physicians, or early twentieth-century German serologists. As the examples suggest, such collectives vary considerably in size, scope and duration, and any individual is likely to be a member of several of them. The latter point is important in explaining how original ideas or ideas that run counter to the prevailing thought style of a collective can arise. Part of the answer is that novel ideas are not simply adjusted to fit *prior* convictions but that all the components of a belief system, including individual beliefs, general background assumptions and ongoing perceptual and behavioural practices, are *mutually* adjusted to maintain their overall coherence. This means that anything and everything – beliefs, assumptions, perceptions or practices – can change while the harmony of the system as such is maintained. Contrary, then, to common misunderstandings and charges, accounts of cognitive process in terms of conceptual systems and thought collectives do not imply either the imprisonment of thought in static, self-confirming circularity or the churning out of robotic individuals doomed to social conformity. Even in the most confined and custom-bound societies, individuals are members of multiple collectives. Families and peer-groups, if nothing else, offer divergent thought styles, to which may be added, in more complex societies, schools, religious sects, political parties and professional groups. Intellectual innovation arises continuously from the ongoing communication of ideas by individuals moving between different collectives (for example, different scientific disciplines); intellectual rebellion can arise from an individual's experience of a significant clash of beliefs or assumptions between different collectives (for example, peer group and family) in overlapping domains (for example, views of class, race or gender relations).

Fleck describes the organisation and social dynamics of thought collectives in considerable detail, picturing each as a set of nested, mutually interacting circles. In many respects, such collectives operate like medieval guilds. At the centre is a small inner ('esoteric') circle of the elite – experts and elders, master builders and laboratory directors. At the periphery is a large ('exoteric') circle – fans, audiences, lay people and the general public. In between is a graduated hierarchy of initiates: students and amateurs, assistants and apprentices. Popular knowledge is, of course, affected (or, as usually seen, 'informed' and 'enlightened') by expert knowledge, but, in Fleck's account, the converse is also true. Esoteric thought – for example, what is framed as scientific knowledge – itself draws on popular beliefs and, to be accepted in the collective, must be attuned to commonsense assumptions even if not simply a confirmation of them. The reciprocal dependence and continuous mutual interaction between outer and inner circles – experts and general public, specialised knowledge and generally accepted beliefs – is crucial in Fleck's account. In scientific disciplines as in religious sects or political parties, these features operate to sustain both the coherence of the collective as a social community and the stability of the shared thought style. There are, however, important differences in the details of these operations, with significant social and intellectual consequences. Thus, where the organisation is strongly hierarchical and the inner circle dominates (as in most religious collectives, Fleck notes), the thought style remains rigid and dogmatic. Conversely, where the organisation is democratic and the degree of dominance between the outer and inner circles remains balanced, thought develops responsively or, one might say, progressively. The best example of such conditions, Fleck observes, is 'modern natural science'.

Elaborating what he calls the specific 'mood' or ethos *(Stimmung)* of modern science, Fleck describes its characteristics, beginning with its expression as 'a shared reverence for an ideal – the ideal of objective truth, clarity, and accuracy':

> It consists in the *belief* that what is being revered can be achieved only in the distant, perhaps infinitely distant future; in the *glorification* of dedicating oneself to its service; in a definite *hero worship* and a distinct [tradition].[69]

As he goes on to explain, this mood is created and sustained by a set of social norms in accord with which the initiate learns that individual personality is kept in the background, that modesty and caution are valued, and that all research workers are presumed equal with regard to the acquisition of knowledge. Other important characteristics include 'a reverence for number and form' and, significantly, 'an inclination to objectivize the thought structures that it has created'. The process of objectivisation occurs in stages, starting with the personal statements of individuals and moving increasingly toward depersonalised ideas stated in special technical terms, 'a language estranged from life' that 'guarantees fixed meanings for concepts, rendering them static and absolute'.[70]

Several points may be made here. One is that Fleck's detailed observations of the structure, dynamics and distinctive ethos – norms and characteristic practices – of modern natural science anticipated and/or proved exceptionally fertile for later twentieth-century sociologists and historians of science, both mainstream figures such as Robert Merton,[71] and those instructed in the largely post-/anti-Mertonian practices of the Strong Programme.[72] A second set of points concerns the political temper of Fleck's views of science – or, rather, their lack of such a temper. As is clear, Fleck's descriptions of the ethos and practices of modern natural science are neither idealised nor especially (self-)congratulatory. Thus he sees its ideals of objectivity and clarity as emergent and effective norms, not as valiantly elected virtues. But also, clearly, his descriptions are not 'cynical' or even, for better or worse, 'critical'. Thus he sees the typical 'hero worship', self-effacement, 'reverence for number and form' and arcane language of modern science as, again, emergent traits and effective norms, not delusions to be eradicated, bad habits to be foresworn, or stratagems to be exposed.

These latter points require emphasis in view of common charges regarding the supposed politics of constructivist science studies. Thus feminists and other academic activists who regard science primarily as a tool of patriarchy, racism, imperialism or homophobia (each of which, of course, it has been – even if the terms here, including 'science', are too crude) charge sociologists of science with being insufficiently 'critical' insofar as they merely describe scientific practices without exposing their political complicities and consequences. Correspondingly, philosophers and other commentators who regard natural science as the last, best abode of reason charge constructivist historians and sociologists of science (typically *conflated* with feminists and other scholar-activists) with 'cynicism' precisely insofar as they do observe its political complicities and consequences and fail ultimately to celebrate it.[73] We shall return to these views and charges in the next two chapters.

XII

How did Ludwik Fleck – a research immunologist working in Lvov – arrive at all these original, challenging ideas about epistemology and the philosophy of science? And why, since those ideas were so original and challenging, was it not until half a century later that they became widely known? And why, since they remain so original and challenging and are now well known to a generation of historians and sociologists of science, are they still largely ignored by philosophers of science?[74] Some answers to these questions are suggested by Fleck's intellectual and professional biography, which has implications both for the character – or, indeed, style – of his thought and also for its fortunes in the relevant collectives. Other answers may be sought in the broader cultural, intellectual and professional worlds in which he operated.

Fleck was a practising scientist, not a philosopher or logician, and he was a biologist, not a mathematician or physicist. His knowledge of classical *Erkenntnistheorie* and current philosophy of science was broad, detailed and discriminating, but he had no professional investment in either as a specific discipline. Although he read extensively in the history of medicine, he did not train as a medical historian or historian of science. In these ways and others, his intellectual profile is significantly different from that of other major figures of twentieth-century history and philosophy of science, both his contemporaries, such as Popper and the leaders of the Vienna Circle (Morris Schlick, Otto Neurath, Carnap and so forth), and such later figures as Carl Hempel, Imré Lakatos or Kuhn. Unlike all these, Fleck was an amateur and outsider in epistemology and the history and philosophy of science. But also unlike them, he was professionally familiar with the practices – technical, social and institutional as well as conceptual – of several scientific fields over a period of more than twenty years. Each of these sets of differences cuts two ways.

First, it seems clear that Fleck's training in biology and life-long immersion in medical research shaped both the specific ways he conceived human behaviour and cognition and also his broader thinking about truth and scientific knowledge. Certainly images and examples of organic development were readier to hand for him than they would have been for a physicist or logician. He was also more familiar and presumably more comfortable than they would have been with related ideas of emergence, flux and holism. As we have seen, Fleck's models and descriptions are typically process-centred and he depicts both cognitive and social phenomena repeatedly or indeed obsessively as interconnected, interactive, mutually dependent and mutually attuned – dynamic, organic, harmonious.[75] In a typical evocation of flux,[76] Fleck makes explicit the organic model that governs his view of historical dynamics, including his sense of the recalcitrance of those dynamics to logical description:

> The history of a field of thought cannot be logically reconstructed any more than the history of a scientific event, because it involves vague and indefinable concepts undergoing a process of crystallization. The more detailed and differentiated the historical description . . . the more complex, interrelated, and mutually constitutive its concepts will be seen to be. An organic structure,

> emerging out of the mutual development of its interacting components, they are a tangle when viewed logically.[77]

Relatedly, he notes that older views and metaphors of warfare in regard to disease and the body – 'invasion', 'attack', 'defence', and so forth, along with the idea of specific pathogenic 'causes' (a relic, he maintained, of archaic ideas of demon-possession) – are now recognised as dubious assumptions. '[The] organism,' he writes, 'can no longer be regarded as a self-contained, independent unit with fixed boundaries.' Rather, he continues (citing developments in bacteriology, morphology, genetics and physiology), it can be seen as a 'harmonious life unit' in which 'the activities of the parts are mutually complementary, mutually dependent upon each other, and form a viable whole through their cooperation'.[78] Like the relation between alga and fungus (the two quite different species of organism that constitute a lichen), the relation between host and parasite or between human body and pathogen could be seen as types of *symbiosis*, as could also, Fleck remarks, a community of social animals or an entire ecological unit such as a forest:

> In the light of this concept [that is, symbiosis], man appears as a complex to whose harmonious well-being many bacteria, for instance, are absolutely essential . . . Thus, rather than as an 'invasion', we might better speak of an infectious disease as 'a complicated revolution within a complex life unit'.[79]

In a revealing comment, he adds: 'This idea is not yet clear, for it belongs to the future rather than the present. It is found in present-day biology only by implication, and has yet to be sorted out in detail.'[80] Indeed, the idea is even now both radical and controversial.[81]

Something of the intellectual-professional background of Fleck's organic models can be glimpsed through his account of important conceptual shifts in contemporary immunology, which was, in effect, his own major thought collective. He writes: '[I]nstead of the prejudicial concept of immunity [protection against invasion from outside], we have the general concept of allergy (a changed mode of reception) . . . '[i]nstead of antibodies, we speak of reagins to stress the lack of direction of the effect.' He explains these shifts as follows:

> Many classical concepts of the field of immunology evolved at a time when, under the influence of the great chemical successes in physiology, misguided attempts were made to explain the whole, or almost the whole, of biology in terms of effects produced by chemically defined substances . . . This primitive scheme is now being abandoned . . . We now speak of states and structures rather than substances to express the possibility that a complex chemico-physico-morphological state is responsible for the changed mode

> of reaction, instead of chemically defined substances or their mixtures being the cause.[82]

What Fleck and his fellow immunologists saw as misguided in these earlier efforts were, first, the description and explanation of complex biological transformations in terms of unidirectional causes and the properties of discrete, autonomous entities; and, second, the general inclination to atomism, binarism (toxins/antitoxins, complements/anti-complements and so forth) and reductionism. Clearly, it seems, the models of *biological phenomena* that dominated *immunology* in the early twentieth century – 'under the influence of the successes of chemistry' – and were later rejected in favour of more holistic, dynamic, organic ones share crucial features with the models of *knowledge* that dominated *epistemology* at the same time, 'under the influence', we could say, of the contemporary successes of physics and logic. It was, of course, just those features of conventional epistemology – its atomism, binarism, agent-centredness, reductionism and unilateral-unidirectional conceptions of causality – that Fleck rejected most strenuously in *Genesis and Development*. And, as we have seen, the models of cognition with which he sought to replace them are, like those of his own succeeding generation of immunologists, overwhelmingly holistic, dynamic, distributed and reciprocal.

Some general observations can be made here. One is that certain key features of the specific cognitive style and related habits of perception, description and explanation that Fleck acquired as a biologist and immunologist evidently carried over into his ways of thinking about knowledge and science and shaped both his tastes and aversions in epistemology. A second observation also concerns relations between thought styles, but has broader implications. It is unlikely that the researchers and theorists whose models Fleck rejected (substance-focused chemists, immunologists of a previous generation, logical positivist philosophers of science) would have denied the complexity, variability, interconnectedness and mutual dependence of the phenomena in question – infectious diseases, human cognition, scientific discoveries and so forth. It is, rather, that, whereas they would have seen simplification and idealisation, along with the search for fundamental elements, ultimate causal agents and basic mechanisms, as the best way to go about studying and explaining those phenomena, Fleck saw such approaches as failing to do the phenomena justice or as only obscuring their constitutive complexities. It is as if, for the former (researchers and theorists), giving an adequate account of a phenomenon meant reducing it to its simplest and most basic components while, for Fleck, it meant specifying its connections and dynamics as fully as possible. So described and distinguished, neither thought style or intellectual-scientific approach appears plainly wrong or intrinsically 'misguided'. Each does appear, however, to be precisely a *style* – historically, culturally, professionally and perhaps to varying extents personally, specific. If now, seventy-five years later, we (or some of us) find a Fleckian style more congenial or natural-seeming, it is, in part, because we ourselves – heirs of the cybernetic revolution; practised in the poststructuralist-constructivist critiques of binarism, reductionism and unilinear process; surrounded by daily talk of and perhaps in the midst of daily experience with world-wide-webs, neural networks, actor-network theory, hyperlinks, feedback loops, self-organising complex systems, and so forth – have become so deeply attuned to that style that it has become imperceptible to us as such.

Although an outsider with regard to the philosophy of science, Fleck was by no means an intellectual loner, even in his most radical speculations and formulations. As just indicated, he had the company of fellow immunologists in some crucial intellectual tastes and, as noted in Chapter 2, the company of some of the major social theorists of his era (Georg Simmel, Karl Mannheim and Émile Durkheim among them) in some of his key ideas. And, of course, he was an *insider* with regard to science as such. The consequences of these aspects of his intellectual biography are evident throughout *Genesis and Development*. Thus Fleck's observations regarding the shaping of scientific hypotheses by such non-'evidentiary' factors as laboratory techniques or specific disciplinary assumptions reflect his intimate familiarity with medical and biological research. Similarly, his appreciation of the role of popular beliefs and the significance of cultural and social forces in the history of science reflects his extensive knowledge of the anthropology, sociology and psychology of his era and his fascination with the details of medical history, from Persian anatomy books to priority disputes between German serologists.

The fact that Fleck was a non-philosopher was not altogether a disadvantage for the development of his thought. Compared with those professionally trained in academic philosophy, such as Popper or Carnap, he was not as confined by its conceptual idioms, classic problems or current issues and positions. Moreover, he had no commitment to affirming its value as a specific intellectual enterprise. Like amateurs more generally, he had the advantage of a certain kind of intellectual freedom and, with it, the possibility or even likelihood of a certain kind of radical originality. By the same token, however, and again like the amateur, autodidact or provincial more generally (as a medical scientist in Lvov, Fleck was all three relative to the philosopher-logicians of the Vienna Circle), he ran the risk of intellectual crankiness, irrelevance and, of course, obscurity. One of the reasons Popper's *Logic* could not be ignored by contemporary philosophers of science was that it engaged so thoroughly the classic figures and issues of the field (Kant and Hume, verification and justification, the problem of induction and so forth) and, furthermore, did so from a thoroughly philosophical perspective. Fleck, on the other hand, who regarded 'speculative epistemology' (precisely because *speculative* – that is, non-empirical) as obsolete, pretty much ignores the history of Western philosophy altogether. No less significantly, Popper's *Logic* is attuned – in its categories of concern, field of allusions and conceptual-rhetorical idiom – to the established thought style of Anglo-European academic philosophy. And that, of course, was the intellectual community and institutional platform – a relatively large and, at the time, prominent one, with a large and highly organised audience already in place – from which both his work and also his presence as an intellectual figure were launched. Fleck, however, had no standing whatsoever in that community, which meant, among other things, no philosophy students or academic colleagues to circulate his ideas. No matter how strenuous Popper's challenge to dominant logical positivist assumptions and methods (and the extent of that challenge remains a matter of dispute),[83] he still wrote *as* a philosopher *to* philosophers, whereas no matter how theoretically original, cogently developed, richly illustrated and concretely documented Fleck's ideas may have been (and they were certainly all of these), he wrote as a non-philosopher to an uncertain audience.

Aside from historians of medicine (primarily, at that time, medical practitioners like Fleck), no one had any professional reason or intellectual need to read a book about the history of syphilis, even one provocatively titled *Genesis and Development of a Scientific Fact* – at least not in 1935.[84] What we think of now as the history and sociology of science were not yet specialised fields and, of course, there was no 'science studies' at all. When Kuhn, then a young American physics-student-turned-historian-of-science at Harvard, picked up Fleck's book some fifteen years later, it was in a quite different institutional and intellectual context – Cambridge, Massachusetts, not Vienna; post-Second World War, not its eve. Given the directions of Kuhn's own thinking at the time and his sense of the energy and promise of his newly chosen field, he must have found in *Genesis and Development* just what he needed, that is, both empirical corroboration and intellectual company – which seems to have been why, as he suggests in the 1962 preface to *Structure*, he decided to follow up Reichenbach's citation of Fleck's book. Fateful footnote.

One further aspect of the context of Fleck's reception should be noted here, namely, the growing division and to varying extents antagonism between the empirical sciences and classic humanistic disciplines, both as cognitive styles and as institutional projects. In accord with the distinction between *Naturwissenschaften* and *Geisteswissenschaften*, theorised as such at the end of the nineteenth century and already quite sharp by the 1930s,[85] academic philosophy increasingly defined its missions and achievements in contradistinction from those of the natural and social sciences. Thus, although Anglo-American analytic epistemology, philosophy of mind and philosophy of science have wished to be seen as *allied to* the natural sciences in valuing rigour, formulating strict definitions and distinctions, and seeking to arrive at universally valid claims, they have not professed to *be* sciences among the other scientific disciplines. On the contrary, analytic philosophers commonly stress the definitively normative, foundational missions of philosophy as such, along with its rationalist commitments and strictly conceptual-analytic, non-empirical methods.[86] Of particular interest in seeking to understand Fleck's reception was the strenuous self-separation of both phenomenology and logical analysis from what was described in Chapter 1 as the heresy of 'psychologism' – that is, the idea that questions of mind, thought, reasoning or knowledge might be examined as psychological and thus, as it was feared or charged, merely subjective or merely empirical phenomena. In such a context, *Genesis and Development*, insofar as it became known at all in philosophical quarters, was likely to have been seen as too tied to empirical fields – psychology, sociology and the history and practice of medicine – to be, as philosophers (still) say, 'philosophically interesting'. Not surprisingly, then, it would be largely in other quarters and not until half a century later that its intellectual interest and power would be recognised.

XIII

We may return now to our general topic for some concluding reflections. With respect to the idea of truth, the central implication of Fleckian and post-Fleckian constructivist epistemology is that, although something *like* 'correspondence' is involved in the situations to which we give that name, it is not a matter of an *objective match* between, on the one hand, statements, beliefs, descriptions or models and, on the other hand, a fixed reality, but, rather, a matter of the production and experience of an *effective coordination* among statements, beliefs, assumptions, observations, practices and projects, all of which are independently mutable but mutually responsive. As pragmatists have always maintained, 'working well' is a key test of the theories or beliefs we call true. The testing, however, is not a discrete act of overt assessment directed toward *already formed* theories or beliefs but a tacit part of the very process of their *being formed* under the specific conditions of their conceptual elaboration, social communication and technological application. Contrary, then, to familiar charges, the pragmatist association of truth with effectiveness does not imply or permit the identification of truth either with personal convenience or with serving the ideological interests of some group.[87]

Given the prevalence of related charges and misunderstandings, I would stress two other final points here. First, in constructivist accounts of the emergence of truth or reality, our beliefs, theories and statements are seen as continuously coordinated with our *bodily* actions and *material* manipulations. Thus the emergent features of the Wassermann Reaction are seen, in Fleck's narrative, as having been crucially shaped by, among other things, the effects of various laboratory manipulations: the decision by scientists or technicians to heat blood samples to just this or that temperature; their cultivating colonies of bacteria in just this or that medium; their adding just this or that quantity of a chemical reagent at a certain point in the procedure; and so forth. This crucially constructivist idea of the reciprocal shaping and coordination of actions, perceptions and theories is quite different, of course, from the more familiar idea that theories are duly corrected, verified or falsified by observation, experience or being tested against Nature. Constructivism is quite distinct, in other words, from classical or positivist *empiricism*. Nevertheless, as should be clear, nothing in these accounts is conceived as occurring in, or issuing from, just people's heads. Accordingly, nothing in them can be accurately described as *idealism* – at least not so conceived – either.[88]

Second, constructivists do not contend that what *makes* a statement or belief true is either just its conformity to some conceptual system, thought style or paradigm, or just its being *thought* true by the members of some thought collective, 'culture' or 'interpretive community'. These can certainly be seen as significant features of many of the situations that we name truth, but – and the difference here is crucial – such conformity or communal acceptance do not *bestow* the putative 'property' of truth on statements or beliefs that would otherwise be, in some putative logical space, false. Such fundamental misunderstandings and consequent charges of absurdity (or solemn assurances, such as that by Susan Haack, that Richard Rorty and other 'postmodernists' have simply *confused* 'the true with what passes for true')[89] are products of the tacit, often complacent, assumption of the very concepts and distinctions that constructivist epistemology questions and rejects or reconceptualises.

In view of the multiplicity of contemporary epistemic communities and the sharp divergences of thought styles, language games and regimes of truth among them, it seems unlikely that any *one* view of truth – realist, positivist, pragmatist or any other – will be universally established as true, at least not very soon. Currently, constructivist views of truth, knowledge, reality and related concepts are proving interesting and useful to scholars, scientists and theorists working in a broad range of fields, including artificial intelligence, cognitive science, theoretical biology, medical anthropology, science education and the psychology of perception – as well, of course, as science studies, cultural studies and literary theory.[90] In finding constructivist accounts conceptually congenial and working with them in their own fields, these scholars, scientists and theorists are not, I think, exhibiting incompetence, irrationality, irresponsibility or a disdain for truth. Nor, the evidence suggests, do such interests, efforts or tastes keep one from being a dedicated research scientist, or an honest, hard-working scholar, or an energetic and effective political activist – if any of these is otherwise one's aptitude or inclination. The worst that a sympathy with constructivist ideas might do right now is threaten an otherwise promising career in analytic philosophy. But it might not do that either.[91]

XIV

A final biographical note on Fleck. A few years after the original publication of *Genesis and Development*, Fleck – a Polish Jew – was arrested by the Nazis and sent first to Auschwitz, then to Buchenwald. Forced at both camps to make typhus vaccine for the German army, he survived the war, emigrated to Israel, and continued up to his death in 1961 to pursue his work both as a microbiologist and as a theorist of science, truth and reality.[92] Fleck's wartime experiences did not appear to change his basic views on any of these topics. He remained an admirer of what he saw as the essentially democratic and intellectually progressive ethos of science; and he remained what we must call, it seems, a radical relativist.

Notes

1. Examples of such views are commonplace. For the charge of incompetence in the humanities, see, for example, Gross and Levitt, *Higher Superstition: The Academic Left and its Quarrels with Science*, pp. 71–106; Sokal and Bricmont, *Fashionable Nonsense: Postmodern Intellectuals' Abuse of Science*; Koertge (ed.), *A House Built on Sand: Exposing Postmodernist*

Myths about Science, pp. 3–6, 9–31, 59–70, 272–85. For the charge of intellectual and moral irresponsibility, see, for example, Himmelfarb, *On Looking into the Abyss*, p. xii. For the idea that post-Reformation intellectual developments in the West make life easier for liars, see Fernández-Armesto, *Truth*, pp. 194–5. For charges of political complicity by 'postmodernists', see Eagleton, *The Ideology of the Aesthetic*, pp. 378–9; and Norris, *Uncritical Theory: Postmodernism, Intellectuals and the Gulf War*. In the past decade, a number of British philosophers have undertaken defences of truth against alleged contemporary denials of it, with Richard Rorty usually indicated as chief miscreant. See, for example, Diamond, 'Truth: Defenders, Debunkers, Despisers'; Haack, 'Concern for Truth: What It Means, Why It Matters'; B. Williams, *Truth and Truthfulness: An Essay in Genealogy*. Although they differ in scope and sophistication, all appeal to some undefined, presumably unproblematic, commonsense notion of truth, stress the ethical and political value of truth (at least when invoked by the right people, with the right motives, under the right circumstances), and deplore the damage to ethics and the polity wrought or threatened by the (alleged) views of Rorty et al.

2. Plato offered a correspondence theory of truth; Hume raised the familiar objection to it; the argument was internal to early twentieth-century logical positivism. For a seasoned summary of the major philosophical positions, see Davidson, 'Epistemology and Truth'.
3. Fleck, *Genesis*, p. 100. In quoting Fleck from the 1979 translated edition, I have modified the translation at various points to restore the emphasis of the original or to provide more precise and/or current counterparts for Fleck's often unusual terms and formulations. Passages where such modifications appear are indicated by the abbreviation 'tm' following the page citations, the first of which refers to Fleck, *Genesis* (1979), the second to Fleck, *Entstehung* ([1935]1980).
4. Fleck, *Genesis*, p. 100.
5. Ibid., p. 101/132, tm.
6. Ibid., p. 101/132–3, tm.
7. Ibid., p. 79. A more literal translation would be: 'A network in continuous fluctuation: it is called reality or truth.'
8. Popper, *Logic*, p. 11.
9. Fleck, *Genesis*, p. 28.
10. The process is often compared, by a dubious analogy, to Darwinian natural selection. See, for example, Kitcher, *The Advancement of Science: Science without Legend, Objectivity without Illusions*, pp. 155–7, and discussion in B. H. Smith, *Belief and Resistance*, pp. 139–40. For Fleck's properly Darwinian analogy, see the passage cited below.
11. Fleck, *Genesis*, pp. 25–6/37–8, tm.
12. A parallel passage appears in Heidegger, 'The Age of the World Picture', written at around the same time:

> It makes no sense whatever to suppose that modern science is more exact than that of antiquity. Neither can we say that the Galilean doctrine of freely falling bodies is true and that Aristotle's teaching, that light bodies strive upwards, is false; for the Greek understanding of the essence of

> body and place and of the relation between the two rests upon a different interpretation of beings and hence conditions a correspondingly different kind of seeing and questioning of natural events. (Heidegger, *The Question Concerning Technology*, p. 117)

Heidegger does not extend the idea as generally as Fleck or pursue its implications in the latter's naturalistic mode.

13. Kuhn reports reading *Entstehung und Entwicklung* in 1949 while a Junior Fellow at Harvard and, in the original preface to *Structure* (Kuhn, *Structure*, pp. xiii–ix), cites it as formative. In his foreword to the 1979 translated edition of *Genesis and Development*, however, he gives a more restrained account of the relation of his own thought to Fleck's (Kuhn, 'Foreword'). For an extensive comparison of the two works, see Babich, 'From Fleck's *Denkstil* to Kuhn's Paradigm: Conceptual Schemes and Incommensurability'. For observations on Kuhn's reticence regarding his intellectual influences, see Nickles, 'Normal Science: From Logic to Case-Based and Model-Based Reasoning', p. 171n6.
14. For the Strong Programme, see Bloor, *Knowledge and Social Imagery*. For an indication of the diverse influences, see Barnes, *T. S. Kuhn and Social Science* and Bloor, *Wittgenstein: A Social Theory of Knowledge*. For an account of Fleck's thought in relation to later constructivist history of science, see Golinski, *Making Natural Knowledge*, pp. 32–5.
15. See, esp., Latour, *Science in Action* and *The Pasteurization of France*. Michel Serres, Gilles Deleuze and Foucault are among evident influences or explicitly indicated inspirations.
16. Not everything said on truth or science in the name of 'constructivism' can be explicated along the lines indicated in this chapter and much said (on truth and science, as on other topics) in the name of 'postmodernism' or 'social constructionism' is no doubt foolish or otherwise problematic. The range of labels and positions here, however, does not explain or justify the extravagant characterisations found in the works cited above (see Note 1), characterisations that seem, in many cases, to be based on statements extracted in just the way described. See, for example, the dismissal of social studies of science in B. Williams, *Truth and Truthfulness*, pp. 2–3.
17. A controversy between Wassermann and an associate over which of them was most responsible for the discovery, published some fifteen years after the events at issue, provides many of the details of Fleck's narrative as well as evidence for Fleck's claim of a strong tendency to retrospective rationalisation in personal biography and intellectual historiography.
18. Fleck, *Genesis*, p. 76.
19. See Feyerabend, *Against Method*, p. 14.
20. Fleck, *Genesis*, p. 76/102, tm.
21. Fleck anticipates an objection to this conception of epistemology via the familiar distinction between the (mere) (context of) *discovery* of a scientific theory, seen as suitable for study by historians, biographers, sociologists and psychologists, and its (ultimate) (context of) *justification*, seen as requiring the skills of logicians and philosophers. His reply is that justification, though important, is internal to the process of science itself and that epistemology

is improperly confined to the post hoc scrutiny [*Überprüfung*] and putative logical assessment of scientific theories (see *Genesis*, pp. 22–3).

22. See, for example, Knorr-Cetina, *The Manufacture of Knowledge*; Pickering, *Constructing Quarks*; Latour and Woolgar, *Laboratory Life: The [Social] Construction of Scientific Facts*.
23. See, for example, Fodor, *Representations: Philosophical Essays on the Foundations of Cognitive Science*, and *Concepts: Where Cognitive Science Went Wrong*; Bechtel, *Philosophy of Mind: An Overview for Cognitive Science*.
24. For surveys and discussion of these developments, see Hendriks-Jansen, *Catching Ourselves in the Act: Situated Activity, Interactive Emergence, Evolution and Human Thought*; Longino, *The Fate of Knowledge*.
25. American analytic philosopher W. V. Quine proposed that, to save itself from irrelevance, epistemology must become 'naturalized', meaning informed and transformed by empirical psychology (Quine, 'Epistemology Naturalized'). Rooney, 'Putting Naturalized Epistemology to Work', discusses the quite limited pursuit of Quine's project in contemporary academic philosophy.
26. Fleck, *Genesis*, p. 27/40, tm.
27. For comparable later studies in social psychology, see Festinger, *A Theory of Cognitive Dissonance*; Nisbett and Ross, *Human Inference: Strategies and Shortcomings of Social Judgment*, pp. 175–9 (on 'belief perseverance'). I discuss these tendencies under the term 'cognitive conservatism' in B. H. Smith, *Belief and Resistance*, pp. 50–1, 84–5, 135, 145, 147.
28. Fleck, *Genesis*, p. 28.
29. Ibid., p. 29/43, tm.
30. Ibid., p. 30/43, tm.
31. Ibid., p. 32/46, tm.
32. In later philosophy of science, this becomes the argument for 'scientific realism' (that is, the view that what science reveals – entities, objects, relations, mechanisms and so forth – is reality itself) from the supposedly otherwise inexplicable success of scientific theories in making predictions. For a good discussion of the argument and Fleckian-style exposure of the fallacy involved, see Magnus and Callendar, 'Realist Ennui and the Base Rate Fallacy'.
33. For more recent views of the anatomical saga, see Laqueur, *Making Sex: Body and Gender from the Greeks to Freud*; Shiebinger, *Nature's Body: Gender in the Making of Modern Science*; Stolberg, 'A Woman Down to Her Bones: The Anatomy of Sexual Difference in the Sixteenth and Early Seventeenth Centuries'.
34. Fleck, *Genesis*, p. 35/48, tm, italics added.
35. Ibid., p. 84/111, tm, italics in text.
36. Ibid., p. 177n3/121n3, tm.
37. Ibid., Fleck cites Carnap, *Die physicalische Sprache als Universalsprache der Wissenschaft* (1931), later translated into English as *The Unity of Science*.
38. Ibid., p. 180n7.
39. Fleck, *Genesis*, p. 181n7. Heidegger's *Sein und Zeit*, aspects of which this passage may recall, appeared in 1927. There is no evidence that Fleck or Heidegger knew each other's work. (See also Note 12 above.)

40. Ibid., p. 92/121–2, tm.
41. Ibid., p. 93/122, tm.
42. Ibid., p. 102.
43. Ibid., p. 93.
44. Ibid., p. 86/113, tm.
45. Fleck was probably not thinking specifically of jazz improvisation in his metaphor of mutual attunement, but the analogies would hold for any form of collaborative musical performance.
46. The holism of Fleck's view of cognition is quite explicit:

> The tenacity of systems of beliefs shows that, to some extent, they must be regarded as units, as independent, style-permeated structures. They are not mere aggregates of partial propositions, but harmonious holistic units exhibiting the particular stylistic properties which determine and condition every single function of cognition. The self-contained nature of the system, the mutual interactions among what is already known, what is to be known, and the knowers [*die Wechselwirkungen zwischen dem Erkannten, dem zu Erkennenden und den Erkennenden*], secures the inner harmony of the system but simultaneously preserves the harmony of illusions, which is quite secure [*auf keine Weise aufzulösen ist*] within the confines of a given thought style. (Fleck, *Genesis*, p. 38/53, tm)

47. Kuhn, 'Foreword', p. ix.
48. See Greenwood, *The Disappearance of the Social in American Social Psychology*, pp. 109–35, for the intellectual context of Kuhn's revulsion and the more general rejection of the idea of socially conditioned cognition in the decades following the Second World War.
49. Le Bon, *La Psychologie des foules* (1895), trans. into German as *Psychologie der Massen* (2nd edn, Leipzig, 1912). Fleck rejects Le Bon's own analyses ('Le Bon . . . sees in any socialization merely a degradation of [individual] psychological qualities' [*Genesis*, p. 180n7]) as well as those of McDougall, *The Group Mind* (1920) and Freud, *Massenpsychologie and Ich-Analyse* (1921). McDougall is rejected for treating the mass as an individual, Freud for seeing all the members as identical and their characteristic activity as following a leader who serves as a common ego-ideal.
50. Fleck, *Genesis*, pp. 179–180n7/146n7, tm.
51. Kuhn, *Structure*, pp. 85, 111–14, 150.
52. See esp. Scheffler, *Science and Subjectivity*.
53. See Kuhn, 'Postscript' (1969), in *Structure*, pp. 198–200, and, twenty years later, his 'Response to Commentators'. For discussion of these changes, see Hoyningen-Huene, *Reconstructing Scientific Revolutions: Thomas S. Kuhn's Philosophy of Science*, pp. 204–5.
54. Pertinently here, one may recall the Tuskegee Syphilis Study, initiated in 1932 and conducted by medical scientists in Fleck's own field, in which a group of African-American men with syphilis continued to be studied for forty years without being provided with or informed of known remedies for the disease (see Jones, *Bad Blood: The Tuskegee Syphilis Experiment*). There are, of course, numerous other examples. For some of them, see Harding (ed.), *The 'Racial' Economy of Science: Toward a Democratic Future*; Proctor, *Value-Free Science?*

55. Fleck, *Genesis*, p. 38.
56. Ibid., p. 39/54, tm.
57. Ibid., p. 39/54, tm.
58. Ibid., tm.
59. Ibid., tm.
60. Ibid., p. 40/55–6, tm.
61. Fleck specifically rejects Ernst Mach's conventionalist view of scientific knowledge (*Genesis*, pp. 9–10).
62. The latter part of this second point – the denial of an equality of validity – is implied by Fleck's argument.
63. Latour, *We Have Never Been Modern*, p. 113.
64. Ibid.
65. Latour's 'absolute relativist', like the 'postmodern relativist' encountered in Chapter 2, is something of a chimera, his or her position being made up of oversimplified versions of the views of constructivist theorists such as Feyerabend, Bloor and Latour himself, views falsely attributed to such theorists by intellectually conservative philosophers and their followers, and, no doubt, views actually pronounced by some sophomores. Similarly, the 'universal' to which Latour says we shall be brought back by a due recognition of the ubiquity and inescapability of networks is quite remote from the classically conceived universal typically defended by such philosophers. Latour's flirtation here with traditionalist positions may reflect his effort to distance himself from fellow sociologists of science, perhaps to avoid guilt by association and thus automatic dismissal, perhaps to imply a position above the fray. See B. H. Smith, *Belief and Resistance*, pp. 137–8, for further discussion of Latour's argument here.
66. See, for example, Foucault, *The Archaeology of Knowledge*, p. 6.
67. See esp. Latour, *Pasteurization*, pp. 79–108, 261–2nn19–21. The account in these pages seems to draw on both Fleck and Foucault, though neither is cited explicitly.
68. Given Foucault's ethical and political concerns and his familiarity with the history and contemporary practices of medical psychiatry and psychoanalysis, he had good reason to be sceptical of those fields. The sorts of research in experimental psychology on which Fleck and Kuhn drew (for example, studies of perception and visual illusions) seem to have been unknown or uninteresting to him, but Foucault was not alone in this ignorance or indifference among mid-century French intellectuals, especially those trained in the phenomenological tradition with its legacy of anti-'psychologism'. On the legacy, see the discussions in Chapter 1 and below.
69. Fleck, *Genesis*, p. 142/187–8, tm, italics in text.
70. Ibid., p. 144/188–9, tm.
71. Merton's studies of the operation of the communal norms of natural science in the 1950s and 1960s independently duplicate various of Fleck's observations (Merton was aware of Fleck's work at the time only through Kuhn's reference to it in *Structure*), and he subsequently sponsored, with Thaddeus J. Trenn, the 1979 English-language edition of *Genesis and Development*. Like Kuhn, however, Merton deplored later twentieth-century sociology of scientific knowledge, which he saw as raw 'subjectivism' and 'relativism' (see Gieryn, 'Eloge: Robert K. Merton, 1910–2003').

72. Fleck's account of the stages of objectivisation anticipates elements of the 'textual technology' of 'virtual witnessing' described in Shapin and Schaffer, *Leviathan and the Air Pump: Hobbes, Boyle and the Experimental Life*, and the movement 'from weaker to stronger rhetoric' traced in Latour, *Science in Action*.
73. See, for example, B. Williams, *Truth*, and Haack, *Defending Science – Within Reason: Between Scientism and Cynicism*. Williams reads the current lull in the culture and science wars as a sign of 'inert cynicism' among the 'truth-deniers' (*Truth*, p. 3). Haack's 'cynics' consist of 'radical sociologists', 'feminists' and other academics 'unanimous in insisting that the supposed ideal of honest inquiry, respect for evidence, concern for truth, is a kind of illusion, a smokescreen disguising the operations of power, politics, and rhetoric' (*Defending Science*, p. 20).
74. Philosophers of science generally know Fleck, if at all, only by name. Few of them appear to have read *Genesis and Development* and most, in my experience, are surprised to learn of its importance for Kuhn. It is not among the canonical texts in the field and is rarely cited in current publications, even where it would seem directly relevant – for example, there are no references to Fleck in Fuller, *Thomas Kuhn: A Philosophical History for Our Times*.
75. Readers familiar with the agonistic or, as it is sometimes said, 'militaristic' idiom of science studies (Latour is the major figure cited here) may be struck by the proliferation of concepts of harmony in *Genesis and Development*. Most of these concepts, however, are readily assimilated to comparable ideas in contemporary history and sociology of science – for example, to Andrew Pickering's 'mangle of practice', which, like Fleck's 'harmony of illusions' and in spite of the very different metaphor, can be understood as the mutual coordination of conceptual, perceptual and material practices in the formation and stabilisation of scientific knowledge (see Pickering, *The Mangle of Practice*).
76. Fleck observes ironically at one point that epistemologists trained in the natural sciences see 'human thinking – at least ideally, or thinking as it should be – [as] something fixed and absolute [and] empirical facts [as] relative' while philosophers trained in the humanities regard 'facts as fixed and human thought as relative'. 'How typical,' he exclaims, 'for both parties to find the fixed in the areas with which they are most unfamiliar!' And, rather strikingly, he adds: '*Would it not be possible to manage entirely without something fixed?*' (Fleck, *Genesis*, pp. 50–1/69, tm, italics added.)
77. Fleck, *Genesis*, pp. 53–4/72–3, tm.
78. Ibid., p. 60.
79. Ibid., pp. 60–1/81–2, tm.
80. Ibid., p. 62/82, tm.
81. See, for example, Maturana and Varela, *Autopoiesis and Cognition: The Realization of the Living*; Oyama, *The Ontogeny of Information*.
82. Fleck, *Genesis*, pp. 62–3/83, tm.
83. See Laudan, *Beyond Positivism and Relativism*; Hardcastle and Richardson (eds), *Logical Empiricism in North America*.
84. For recent studies of Fleck's work from the perspective of contemporary medical theory, see van den Belt, *Spirochaetes, Serology, and Salvarsan:*

Ludwik Fleck and the Construction of Medical Knowledge about Syphilis, and Brorson, *On the Socio-Cultural Preconditions of Medical Cognition: Studies in Ludwik Fleck's Medical Epistemology*. Both deal extensively with Fleck's constructivist views as such.

85. See Collini, 'Introduction' in *The Two Cultures*.
86. A twist on all this has emerged in the wake of the 'science wars', with the supposed exposure of a 'collapse of standards' in the humanities and, as detailed in Chapters 2 and 5, the supposed takeover of scholarship in various fields by dubious 'postmodernist' doctrines. Accordingly, analytic philosophers appear more eager than ever to ally themselves with the natural sciences, now by way of distancing themselves institutionally from other presumptively dishonoured humanities disciplines.
87. For inferences along such lines in regard to pragmatist-constructivist conceptions of truth, see the works by Eagleton and Norris cited in Note 1, above.
88. Certain elements of constructivist thought are congruent, however, with philosophical views of reality as in some way mind-dependent (on the latter, see Rescher, 'Idealism'). The misunderstandings in question here are exacerbated in some quarters by conflations of the constructivism/realism polarity with the politically freighted idealism/materialism polarity. Accordingly, where 'materialism' is associated (for example, by Marxists) with politically progressive views, constructivism, conflated with a materiality-denying 'idealism', may be under suspicion of politically undesirable or irresponsible implications (for example, subjectivism, solipsism and so forth).
89. Haack, 'Concern for Truth', p. 61.
90. For constructivist views in artificial intelligence and cognitive science, see, for example, Port and van Gelder (eds), *Mind as Motion: Explorations in the Dynamics of Cognition*; in theoretical biology, see, for example, Maturana, 'Reality'; in science education (primarily via the work of developmental psychologist, Jean Piaget), see, for example, von Glasersfeld, *Radical Constructivism*; in the psychology of perception, see, for example, Blake and Yuille (eds), *Active Vision*; in cultural anthropology, see, for example, Gibson and Ingold, *Tools, Language and Cognition in Human Evolution*, pp. 449–72. For recent explicitly Fleckian accounts in medical anthropology and sociology, see, respectively, Young, *The Harmony of Illusions: Inventing Post-Traumatic Stress Disorder*; and Epstein, *Impure Science: AIDS, Activism and the Politics of Knowledge*.
91. See, for example, Hacking, 'The Self-Vindication of the Laboratory Sciences', and *The Social Construction of* What?; Giere, *Science without Laws*; Longino, *The Fate of Knowledge*. These philosophers' sympathy with constructivist views falls short, to various extents, of embracing them.
92. For Fleck's descriptions of his experiences and activities in the camps, see Schnelle, 'Microbiology and Philosophy of Science, Lwów and the German Holocaust: Stations of a Life – Ludwik Fleck 1896–1961'. For a selection of Fleck's translated papers, including his writings after the war, see Cohen and Schnelle, *Cognition and Fact*, pp. 113–60. For further details of Fleck's life and career, see also Trenn and Merton, 'Biographical Sketch'.

Chapter 4

Cutting-Edge Equivocation: Conceptual Moves and Rhetorical Strategies in Contemporary Anti-Epistemology

We can derive some sense of the way intellectual life is experienced in some era from the recurrence of certain metaphors used to describe its conduct – for example, the frequency with which, in our own time, intellectual projects and achievements are described in terms of navigational finesse: the charting of passages between extremes, the steering of middle courses, the avoidance of the twin perils of Scylla and Charybdis. Thus an advertisement for philosopher Susan Haack's book, *Evidence and Inquiry*, features a statement by Hilary Putnam praising the author for 'elaborating and persuasively defending a position . . . which adroitly steers between the Scylla of apriorism and the Charybdis of scientism'.[1] Or again, *Image and Logic*, by historian of science Peter Galison, is commended by its reviewer, professor of physics Michael Riordan, for 'adroitly side-step[ping] one of the most contentious issues at the heart of current science wars . . . whether scientific measurements stand on their own as arbiters of reality, as the positivists insist [o]r, . . . as the relativists counter . . . predominantly reflect the biases of the culture that constructs them'. Riordan concludes the review by applauding Galison for 'tak[ing] a mighty stand in the middle of these debates, a richly philosophical voice of moderation with which both extremes must now reckon'.[2]

There is some question, of course, as to whether Riordan's statement of the issue in the so-called science wars is altogether even-handed and, relatedly, whether his report of the views of whomever he means by 'the relativists' (he alludes in passing to Thomas Kuhn) is itself accurate. One might also raise the question of how such possible bias on Riordan's part might be measured and, in the case of disagreements on such matters, of who or what would stand as *their* arbiters. Indeed, each of these questions reflects more general issues – for example, the limits of observational objectivity and the commensurability of varying conceptions of epistemic value and judgement – that are also currently contentious but, with significant rhetorical effect, not acknowledged here as such. In other words, the very terms in which 'moderation' is praised here promote one side of a conflict (or of several conflicts) over the other(s) and perpetuate dubious conceptions of the issues involved as well as the nature of the alternative positions. In these respects, however, the review is quite typical of the class of moves and strategies I shall be discussing here.

An especially self-conscious description of navigational finesse occurs on the opening pages of a recently published book, *On the Origin of Objects*, by philosopher/computational-theorist Brian Cantwell Smith, who writes as follows:

> This book introduces a new metaphysics – a philosophy of presence – that aims to steer a path between the Scylla of naïve realism and the Charybdis of pure constructivism. The goal is to develop an integral account that retains the essential humility underlying each tradition: a form of epistemic deference to the world that underlies realism, and a respect for the constitutive human involvement in the world that underwrites constructivism . . . [T]he project requires finding . . . a way to feed our undiminished yearning for foundations and grounding, while at the same time avoiding the reductionism and ideological fundamentalism that have so bedeviled prior foundationalist approaches . . . [T]he proposal shows . . . how an irrevocable commitment to pluralism is compatible with the recognition that not all stories are equally good.[3]

As this suggests, however, Smith's navigational feat risks becoming not so much a steering-between as a steering-in-two-directions-at-the-same-time, with the alternate perils – of stasis or shipwreck – that such a project evokes. Like Riordan's moderation-praising review of Galison, Smith's launching of his extremity-avoiding voyage involves a number of question-begging turns. How general, for example, is the set of people who 'yearn' – with or without diminishment – for 'foundations and grounding'? And is it 'reductionism' and 'ideological fundamentalism' that trouble critics of various foundationalisms, or something more consequential for Smith's own project, such as fundamental conceptual incoherence? More significant here than the question-begging, however, is the affirmation of what appear to be contradictory positions. For if one endorses a constructivist understanding of 'human involvement in the world' as *constitutive*, then one cannot consistently retain the 'epistemic deference' to a presumptively *autonomous* reality that generally defines realism. It is this sort of elaborated affirmation of mutually incompatible doctrines or, in the name of middle-road moderation, the simultaneous or rapidly oscillating avowal and disavowal of both traditional and more or less radically revisionist positions that I shall discuss here as 'equivocation'. In regard to a number of currently volatile intellectual issues, it appears to be a major – perhaps predominant – mode of theoretical discourse in our time.

Some signal features of the mode are illustrated by Smith's book, which, in response to evidently intractable conceptual problems in contemporary computational theory, questions the viability of a number of key assumptions and formulations taken over by computer scientists from classical metaphysics and philosophy of mind. In the course of his pursuit of a 'successor metaphysics', Smith outlines an admirably original and, in some respects, unmistakably constructivist epistemology and Latourian or Heraclitean ontology.[4] Anxieties about the dangers or absurdities lurking in such positions, however ('I'm-OK-you're-OK-pluralism', solipsistic idealism and so forth), along with axiomatic commitments to some dubiously privileged intuitions (the conviction of most computer scientists, for example, that human cognition and the operations of artifactual computers reflect the same underlying mechanisms) lead him recurrently back to only superficially reformed versions of the ideas and assumptions he otherwise questions and seeks to escape. Thus while he stresses that objects and their properties are neither fixed nor prior but emerge from dynamic, context-dependent interactions between actively 'registering' subjects and fluid 'object-regions', he also insists that the relationship between subject and object is 'fundamentally asymmetrical': 'When subject and object part company, *the object wins* . . . Subject really is *less*; world, *more* – as much in terms of potency and worth as in terms of content or substance.'[5] Or again, he observes that 'although there is something right about speaking of individual subjects as the entities or agents that register, this is not to deny that in all likelihood it will be whole cultures, language communities, communities of practice, or collectivities of people-and-instruments-and-organizations-and-documents-and-tools-and-other-essential-but-expensive-entities that are the full sustaining locus of this intentional achievement'[6] – which balances a residual Cartesianism against a sophisticated but hedged constructivism, on the toes, so to speak, of a paradoxically individualistic but collectivist intentionalism.

Significantly enough, Smith cites, as elucidating his own project, a passage from Donna Haraway's influential essay, 'Situated Knowledges: The Science Question in Feminism and the Privilege of Partial Perspectives', that reads as follows:

> So, I think my problem and 'our' problem is how to have *simultaneously* an account of radical historical contingency for all knowledge claims and

> knowing subjects, a critical practice for recognizing our own 'semiotic technologies' for making meanings, *and* a no-nonsense commitment to faithful accounts of a 'real' world, one that can be partially shared and friendly to earth-wide projects of finite freedom, adequate material abundance, modest meaning in suffering, and limited happiness.[7]

The Haraway passage will concern us a bit later when it is cited by feminist philosopher Sandra Harding in response to a different set of intellectual pressures. For the moment, however, we might note that it evidently figures for a number of its readers both as a compelling formulation of a key problem of contemporary thought and also as a model for its solution.

At the least, these recurrent figures of navigational manoeuvring – avoiding extremes, finding a middle course between twin perils, holding on to both sides, keeping everything onboard – suggest that, for many scholars, contemporary intellectual life is a stressful venture, fraught with danger and haunted by anxieties about the seductiveness, naïvety or fatality of certain moves, choices or rejections. Part of my interest here is what has made that the case in contemporary epistemology (or anti-epistemology) and how various responses to that situation illuminate the dynamics of intellectual life, both currently and more generally.

I

A word should be said first about the perilous seas within which all this anxious steering is occurring. Clearly there are significant contemporary challenges to classical epistemology and mainstream philosophy of science: new ways of answering classic questions concerning the formation and validation (or is it contingent stabilisation?) of belief, new questions about the nature and operations of scientific knowledge, and new assessments of the role of academic philosophy both in posing such questions and in grounding or adjudicating their answers. These challenges are by no means recent in origin. Some have been part of the philosophical tradition since Protagoras; others can be traced without difficulty to the ideas of Friedrich Nietzsche, Ludwig Wittgenstein, Martin Heidegger, William James and John Dewey; yet others have emerged during the course of this century from research and analysis in the scientific disciplines themselves, for example, in quantum theory and, more recently, in developmental biology and cognitive science. Work in all these fields has indicated the need to review and, to some degree, revise traditional ideas and conventional wisdom – formal and informal – about knowledge, science and cognitive processes.

At the same time theorists and scholars in various relatively new fields, including feminist epistemology and constructivist history and sociology of science, have pressed these challenges with especially aggressive energy and in quarters quite close to home – that is, in academic philosophy itself.[8]

Responses to these developments among philosophers, scientists and scholars or theorists in related fields vary, as might be expected, in relation to individual intellectual history, professional identity and status, cognitive taste and temperament, and other commitments and agendas (for example, religious or political), and range from eager embrace and declarations of close alliance to excoriation and frenzies of refutation. It is in this context that we must understand the general sense of peril and anxiety I have indicated. I turn now to a closer look at some specific expressions of it.

II

The types of equivocating moves and strategies to be discussed here sometimes announce themselves as a middle way between what are described as 'extremes'. It is not uncommon, of course, for diverging intellectual positions to become polarised and certainly not impossible for a proposed via media to offer, and operate as, a genuinely valuable alternative to two manifestly strained sides. In the cases that concern me here, however, the alleged extremes are typically somewhat gerrymandered, one of them being, in fact, the currently orthodox doctrine itself, but in an especially hoary version that, as such, has few if any contemporary advocates (for example, 'pure apriorism', 'naïve realism', or 'dogmatic positivism') while the other alleged extreme is, in fact, the currently most powerful challenger but characterised in terms that make it appear dismissible out-of-hand (for example, 'trendy scientism', 'corrosive hyperrelativism' and so forth). This leaves, to occupy the space of the purportedly moderate middle way, either the orthodox doctrine once more, though described this time in duly reasonable-sounding or updated terms ('moderate realism', 'historically informed post-positivism' and so forth), along with, perhaps, an appliqué of selected but denatured elements of the contemporary alternative or, as in several of the examples considered below, a conceptually unstable amalgamation of crucial but also mutually contradictory elements of each.

Conceptual instability is often a key problem here. Innovative theoretical proposals, including radically innovative ones, inevitably preserve some elements of traditional thought, typically reworking or redefining them in conjunction with significantly novel elements or extending them, thus conjoined, into significantly new domains of application. Indeed, though not usually motivated by anxious extremity avoidance, such fertile combinations of old and new are probably the most common forms of conceptual innovation and creative transformation in any field. A major problem with the fundamentally equivocating hybrids described above, however, composed as they are of attenuated and/or patched-together *mutually cancelling* concepts, is that they can do little theoretical work and, indeed, commonly cannot be extended beyond the pages on which they are framed, even by their admirers. (It is not clear, for example, that Haraway's elaboration, in 'Situated Knowledges', of the idea and project framed in the much-cited passage quoted above – that is, that her and our problem is how to have, *simultaneously*, a large number of evidently desirable or necessary-seeming but possibly incompatible things – amounts to much more than a series of reiterated affirmations of that idea and project themselves.) Conversely, what gives many of the 'extreme' proposals their conceptual power is, among other things, precisely their *extremity* – that is, the unhedged explicitness of their questioning or rejection of various traditional ideas and the consistency of the alternative ideas they develop. Contrary to what the term 'extreme' may suggest, these intellectual virtues are the product not of uncontrolled excess or exhibitionist derring-do but, rather, of an effort at clear and precise formulation and a rigorous working-through of theoretical and practical implications – at least where such characteristics are in fact displayed. The intellectual virtues of some challenges to orthodoxy – 'extreme' and otherwise – may, of course, be quite meagre.

Some specific examples will be useful here. Within the past few years, a number of works have appeared offering either to mediate between or to synthesise traditional philosophy of mind and such relatively new fields as artificial intelligence, cognitive science, constructivist epistemology and/or dynamical systems theory. Such projects confront a number of rhetorical difficulties: not only the familiar problems involved in making novel, technically complex ideas comprehensible to nonspecialists but also, more significantly here, the difficulties that the authors of such works may encounter in articulating their own positions with regard to ideas currently orthodox or heretical in their own fields and, relatedly, the task, often cognitively stressful as well as professionally delicate, of negotiating their own intellectual allegiances or even identities with regard to their home disciplines. One sign of these difficulties is the emphasis in such works, often in tandem with the announcement of radical transformations, fundamentally new paradigms and significant alternatives to traditional thought,[9] on the need to *retain* certain egregiously traditional ideas, an emphasis that becomes quite problematic when it is the viability or necessity of *just those ideas* that is disputed most strenuously by the most controversial but arguably fertile developments in the relevant new fields.

When we focus on the stated reasons for preserving this or that allegedly indispensable element of traditional thought, we begin to see the sorts of pressures that lead to these cognitive stresses and rhetorical evasions. One such reason is that the idea or method in question is, as it may be said, so intuitively compelling or well established in the field that it cannot and should not be abandoned. Thus, in his mediating-synthesising work, *Being There: Putting Brain, Body, and World Together Again*, philosopher/cognitive theorist Andy Clark, explaining the need to retain traditional ideas of internal representation, computation and 'stored programs' in the brain despite the ('radical') effort by other theorists to model the dynamics of cognition without appeal to such ideas, argues that 'it would be folly to simply jettison the hard-won bedrock of cognitive scientific understanding that involves [such] ideas'.[10] This seems to say, however, only that those ideas are very orthodox indeed. Clark's rather amusingly (and multiply) mixed metaphor here, that is, the supposed folly of jettisoning (hard-won) bedrock, invites a pertinent query: Would it be better, navigationally speaking, to keep bedrock onboard or, of course, even to *begin* a voyage with such cargo?[11]

Significantly, Clark argues in a more recent essay for a policy of 'accommodation not elimination' in negotiating the differences between the currently prevailing computational/representational model of cognition and the alternative dynamical/ecological model proposed by other theorists.[12] Metaphors of negotiation or accommodation generally sound more reasonable in responses to intellectual rivals than those of outright warfare or total elimination and, where otherwise appropriate, are certainly preferable to denunciation, demonisation and other rhetorical strong-arm tactics. As in macropolitics, however, so also in the micropolitics of intellectual struggle: a refusal to acknowledge the strength of a challenger or the extent and possibly radical nature of the difference between an opponent's views and one's own is likely to be ultimately debilitating. The question in regard to Clark's efforts here (and comparable ones at theoretical synthesis elsewhere) is whether the two now hopefully reconciled theories are, in fact, ultimately compatible and, specifically here, whether just piggybacking elements of the dynamical/ecological model of cognition onto the otherwise unmodified representationalist/computational model solves the crucial conceptual problems in the latter that led to its rejection – and to the related development of various alternative accounts – in the first place.[13]

Another – perhaps the major – reason commonly offered for retaining one or another traditional idea is that its rejection would amount to an embrace of what is seen as the *only* and a clearly foolish (for example, solipsistic or self-refuting) or dangerous (for example, leading to Auschwitz or endorsing clitoridectomy) alternative. Thus explaining why, versus 'pure constructivism', a realist conception of 'a world out there' must be retained in any adequate metaphysics, Smith observes: 'There is more to the world than us . . . more than our imaginations, more than our experience, more than our thoughts and dreams.'[14] Contrary, however, to the implications of Smith's assurance, the idea that the world is *nothing* more than our imaginations, thoughts, dreams or, in any idiomatic sense, experience is by no means the only alternative to realism, as one discovers if one examines various nonrealist epistemologies in their actual, as distinct from distorted or sloganised, articulations: for example, in Ludwik Fleck's classic (and arguably 'pure [proto-]constructivist') *Genesis and Development of a Scientific Fact*.[15]

The configuration of mutually sustaining but individually dubious assumptions, anxieties and charges just indicated is a recurrent feature of contemporary theoretical controversy and, as such, interesting along a number of lines. The point I would stress here is that, as in the example just examined, the scandalising alternative to orthodox thought commonly said to require the reaffirmation of traditional ideas is often a quite empty position: maintained *as* a 'position' by nobody at all but functioning crucially within the intellectual tradition as a self-haunting, self-policing, self-perpetuating 'other', continuously regenerating, by sheer contradistinction, the substance of the tradition's self-defining orthodoxies. Thus cognitive psychology is kept in line by a straw-man behaviourism, theoretical biology polices itself in opposition to a shadowy Lamarckism,[16] and orthodox epistemology is haunted (and kept orthodox) by the phantom heresy of 'relativism'.

Ghostly or empty though these alternatives may be, their invocation is nonetheless powerful. To say that some set of views is orthodox is to say, among other things, that it is institutionally well established and thus part of the conventional training and ongoing discursive and conceptual operations of some field or discipline. For many people rigorously trained in the field in question and especially for those currently operating in it professionally, it may be very difficult to think otherwise and, in a way, unnecessary for them to do so, at least as long as they remain within the institutional orbits of that orthodoxy. For others in that field, perhaps no less rigorously trained but operating at its margins (interacting, perhaps, with scholars, scientists or theorists in other disciplines or working in relatively peripheral areas), some elements of the constitutive heresy may appear, especially in contemporary articulations, intellectually compelling and appropriable but the possibility of affirming or incorporating them explicitly may be inhibited by considerable anxiety, including a well-enough instructed fear of social and professional punishment – for example, scorn or ostracism – at the hands of academic associates or disciplinary colleagues. We shall see some vivid examples of such dynamics below.

III

In addition to equivocation in a variety of technical and looser senses, there are several other moves and strategies of related interest here, including what I call ritual exorcism or blackening the devil – that is, the vigorous dissociation of one's own manifestly (explicitly acknowledged or indeed stressed) non- or anti-orthodox ideas from one or another officially heretical position, accompanied by the voluble badmouthing of the position thus named. Thus the pragmatist (and arguably relativistic) philosopher Richard Rorty strenuously rejects what he calls 'relativism';[17] the constructivist (and arguably postmodernist) theorist of science Bruno Latour derides what he calls 'postmodernism';[18] and the explicitly post-Cartesian (and arguably behaviourist) neuroscientist Antonio Damasio pauses to disparage behaviourism.[19] Such otherwise gratuitous disavowals reflect many of the same institutional dynamics as does equivocation, with comparable short-run rhetorical advantages but long-run intellectual costs. For insofar as it reinforces the idea that the position thus disparaged (for example, behaviourism) is, in fact, monolithic and either plainly foolish or plainly sinister as commonly believed or that the label thus strenuously rejected (for example, 'relativism' or 'postmodernism') names a position that is actually maintained with the foolish claims and dangerous entailments commonly attributed to it, ritual exorcism strengthens a major line of justification for the continued dominance of the intellectual orthodoxy that is otherwise being explicitly challenged. Such devil-blackenings are also more generally intellectually damaging insofar as they endorse the prejudices and foster the anxieties of less knowledgeable colleagues, students or members of the public and thus – this being the way canonical distortions perpetuate themselves – effectively deprive new generations of scholars and theorists of potentially useful intellectual resources. These and other points touched on here can be illustrated with another set of examples, drawn this time from feminist epistemology or, possibly, anti-epistemology.

In a recent essay explaining the idea of 'standpoint epistemology' and promoting her own related position of 'strong objectivism', feminist philosopher of science Sandra Harding argues that one may maintain that 'all knowledge is socially situated versus the conventional idea that beliefs count as knowledge only when they break free [of] . . . local, historical interests, values, and agendas', but still not 'slide' into 'relativism'. For, she continues, citing the passage by Haraway quoted above, 'it turns out to be possible to have *simultaneously* an account of radical historical contingency for all knowledge claims and knowing subjects . . . and a no-nonsense commitment to faithful accounts of a "real" world'.[20]

It is not insignificant that the word *real* appears in this passage in quote-marks. Feminist epistemologists, caught between the particular social-political commitments that motivate and define their project *as* feminist and the largely universalist assumptions and aspirations that prevail in academic philosophy, are driven repeatedly to equivocating moves and gestures: arguments and formulations that explicitly challenge the defining claims, terms and missions of rationalist-realist epistemology and normative-universalist philosophy of science but that also pointedly affirm various crucial elements of each. Thus, in the Harding-Haraway passage just quoted, while the idea of the '[socially] situated' nature of knowledge is framed as 'versus' the conventional idea of genuine knowledge as transcendent, the assurance of a 'no-nonsense commitment to faithful accounts of a "real" world' affirms an epistemologically orthodox realism *versus* a threatening slide into heterodox relativism *and*, in the same breath, signals a scepticism toward that same orthodoxy via quote-marks on the crucial term *real*.

If feminist (anti-)epistemologists such as Harding and Haraway feel called upon repeatedly to affirm the reality of Reality and the possibility of faithful accounts of it, it is largely because of the common conviction and frequent charge by academic philosophers (and those whom they have instructed) that to maintain the radical contingency of all knowledge-claims is ipso facto to deny the possibility of – in their idiomatic/informal as well as technical/formal senses – true, accurate or objective reports. Not all the pressure for such affirmations of intellectual orthodoxy comes, however, from the philosophical side of the feminist/epistemologist double-bind. Related political pressures originate in familiar Marxist-feminist distinctions, such as that between genuine understanding and mere ideology or 'false consciousness', and also in such specifically feminist projects as the 'legitimation' (as it is termed) of women's disputed accounts of their own experiences – for example, of rape. It is widely believed that the political success of such projects depends on the rhetorical/justificatory force of such distinctions, and that the latter depends in turn on the possibility of invoking a world of *transcendently* objective facts or *universally* valid truth claims – facts and claims, that is, that are emphatically *not* ('merely') socially (or otherwise) 'situated'. Such convictions double (and are, of course, historically derivable from) those of a more traditional realist-rationalist epistemology and, accordingly, double the anxieties that attend the more general project of feminist epistemology.

The gestures elicited by such anxieties are sometimes instructively self-reflexive, as in the essay by feminist philosopher Lorraine Code titled 'Taking Subjectivity into Account'. In the body of her essay Code argues that the paradigmatic form of knowledge in epistemology and related discourses, such as judicial theory, should not be our presumptively but dubiously objective knowledge of *objects* but, rather, our manifestly subjective knowledge of *other people*, specifically the people with whom we have personal relationships. It is not an easy argument to make, and Code's success in making it is limited. My primary concern here, however, is the question she raises toward the end of her essay, where she asks whether her argument stressing the significance of a subjectivist epistemology for feminism means that 'feminist epistemologists must, after all, "come out" as relativists?'[21] Her answer to that question is 'a qualified yes', but it is followed by a significant *yet*: 'Yet the relativism that my argument generates is sufficiently nuanced and sophisticated to escape the scorn – and the anxiety – that "relativism, after all" usually occasions.' Indeed, the relativism that Code generates in the succeeding pages of her essay is so thoroughly nuanced and sophisticated – or, one could say, haunted, hedged, and attenuated ('refus[ing] to occupy the negative side of the traditional absolutism/relativism dichotomy . . . at once realist, rational and sufficiently objective', and so forth)[22] – that one could very well mistake that qualified *yes* for an unqualified *no*.

The terms of Code's self-characterisation raise some immediate questions. First, one may wonder if she believes that the relativism commonly scorned by her disciplinary colleagues and others – the ideas, presumably, of heterodox theorists and philosophers such as Nietzsche, Kuhn, Paul Feyerabend, Nelson Goodman and Rorty – is *not* nuanced or sophisticated. And, if it is not from the work of figures such as these, then one may ask from what crude, rude, or naïve relativism Code is here distancing herself. Such questions arise not only because Code names no names but also because, as pointed out above, there is reason to think that the foolish or dangerous ideas commonly scorned *as* 'relativism' – the idea, for example, that all beliefs or accounts are equally valid (under all conditions, from all perspectives)[23] or that the world can be constructed just as we choose – constitute a phantom heresy, without visible, palpable or citable adherents. The phantom appears to be continuously generated by the seesaw logic of orthodox epistemology itself: if not classic *realism*, then classic *idealism*; if not classic *objectivism*, then classic *subjectivism*; if not *one uniquely valid* interpretation/judgement/theory, then *all equally valid* interpretations/judgements/theories; and so forth. It may be given added apparent substance, however, by such heavy rhetorical scaffolding as glib conflation, crude decontextualisation, dubious imputation, tendentious paraphrase and slapdash intellectual history.[24]

To the extent that this is the case – that is, that the menace/heresy of relativism is substantively empty – it appears that, rather than spending so much time and energy anticipating and attempting to deflect such charges, Code and other feminist (anti-) epistemologists might do better exposing their hollowness, criticising in earnest the entire conceptual systems that continuously generate and sustain them, and exploring the possible value, for feminism and social theory more generally, of the specific, elaborated ideas of the thinkers whose views are commonly so charged and dismissed. Indeed, in view of the quantities of intellectual labour that, in the absence of such direct challenges to demonology, must now go into protecting their own efforts from such out-of-hand dismissals, it would seem to be the more intellectually efficient as well as responsible way to go. As it stands, the anxiety to avert charges of relativism is not only a considerable distraction from other potentially more productive intellectual activities but leads, often enough, to strained and tangled formulations. Code acknowledges as much in her own case in the final paragraph of her essay. 'There are,' she writes, 'many tensions within the strands my sceptical-relativist recommendations try to weave together,' tensions to be expected, she observes, 'at this critical juncture in the articulation of emancipatory epistemological projects'.[25] The question, however, here and more generally, is whether both the emancipatory and the epistemological commitments of such projects, to the extent that either of them is conceived along conventional political or conventional philosophical lines, might not be holding their more radical critical and creative energies hostage.

I return to that question below but wish, first, to pursue a bit further the terms of Code's description of the scorn that motivates feminist anxiety about relativism. She writes:

> The opponents of relativism have been so hostile, so thoroughly scornful in their dismissals, that it is no wonder that feminists, well aware of the folk-historical identification of women with the forces of unreason, should resist

> the very thought that the logic of feminist emancipatory analyses points in that direction . . . The intransigence of material circumstances constantly reminds them that their world-making possibilities are neither unconstrained nor infinite . . . In fact, many feminists are vehement in their resistance to relativism precisely because they suspect – not without reason – that only the supremely powerful and privileged . . . could believe that they can make up the world as they will and practice that supreme tolerance in whose terms all possible constructions of reality are equally worthy.[26]

Code goes on to question the accuracy of the latter suspicion but only to suggest – equally tenuously, I think – that 'only the supremely powerful and privileged' could believe there was but *one* truth.

Many questions could be raised about this argument (for example, are feminists, as Code implies, more familiar with 'the intransigence of material circumstances' than other people who question traditional objectivist thought, and do the latter really need reminding 'that their world-making possibilities are neither unconstrained nor infinite'?), but two points require emphasis here. First, I say *equally* tenuously because both views, those Code attributes to many feminists and her own, reflect the logically and historically dubious assumption that people's epistemologies line up squarely with their social and/or economic situations, which reflects in turn the comparably dubious assumption that the inherent political value of a theoretical position can be determined by the particular political positions of the people who happen, at a given time, to maintain it. The converse view, I would stress, is not that there is *no* relation between people's theoretical preferences and their social or economic situations, or between the political value of a theoretical position and the politics of those who propose and/or promote it, but, rather, that both relations are highly mediated by other variable conditions. These would include other intellectual and/or political commitments of the people in question and other aspects of their social situations and also significant features of the particular intellectual and institutional contexts in which the theoretical preferences in question are maintained or played out. Certainly no generalised ideology-critique or accurate all-time prediction of political uptake could have been produced in the past for such theoretical preferences as polytheism versus monotheism, logical positivism versus Hegelian idealism, or Darwinism versus biblical creationism. Nor, I think, can any be usefully produced for such current theoretical preferences as epistemic pluralism versus epistemic monism or constructivist science studies versus realist epistemology.

The second point concerns the idea, which Code endorses ('many feminists . . . suspect – not without reason'), that there exists a set of people who do, in fact, 'believe that they can make up the world as they will and practice that supreme tolerance in whose terms all possible constructions of reality are equally worthy'. The idea is worth pausing over since it has, as we shall see, considerable circulation among feminist epistemologists as well as other contemporary philosophers but appears quite questionably derived. A likely derivation here as elsewhere is a misunderstanding of the frequently mentioned but evidently not always carefully read 'symmetry postulate' of the Strong Programme in the sociology of science. The postulate, which has nothing to do with tolerance in any of the usual senses of the term, maintains not that all constructions of reality are equally *worthy* but, rather, that the credibility of all constructions of reality, including those now commonly accepted as true or reasonable, should be regarded as equally *needful of explanation* and as explicable, in principle, by the same *general types of causes*.[27] A key methodological point of departure – not epistemic judgement – in contemporary science studies, the symmetry postulate is routinely transformed into a fatuous egalitarianism (everything is equally true, good, worthy, valid and so forth) by those who know it primarily by hearsay (for example, through arguments against it by traditionalist philosophers) or interpret its implications via the seesaw logic described above. The anxiety-eliciting configuration to which Code and other feminists are responding, however, is not merely a common and especially obdurate set of intellectual misunderstandings but a common though dubious set of political assumptions and assessments as well.

The symmetry postulate, correctly stated and appropriately interpreted, implies, among other things, that no belief or knowledge claim can be presumed *intrinsically* credible (or 'valid'). This is an idea, one might think, that would appear highly serviceable to feminists as well as to members of other groups whose political or social subordination is underwritten by what are claimed to be, by those in dominant positions, self-evident facts. And, as it happens, a number of politically concerned scholars *have* found the idea serviceable and put it to use accordingly in their research and analyses.[28] But there's a catch. For, by the same token, the symmetry postulate creates difficulties for feminists (and others) who want to maintain that certain beliefs or knowledge claims *are* intrinsically credible: for example, the truth claims made by women about what they experience as rape or, as in some versions of standpoint epistemology, the knowledge claims made from the perspective of members of one or another marginalised group.[29] What Code and others refer to as *supreme tolerance*, then, might be better described as *supremely even-handed intolerance* – that is, a principled rejection of all claims to generic epistemic privilege, not only the conventionally privileged knowledge claims of scientists *as such* but also claims made for the intrinsically privileged knowledge of women *as such*. For some feminists, that's all they need to know about the symmetry postulate to know that it's bad news politically. There are, however, good reasons for them to look more closely at the political implications of the postulate and also at those of constructivist accounts of knowledge more generally. For although such ideas and accounts may unsettle certain standard rhetorical practices of the political left (as well as the political and intellectual right), they also suggest a broad range of alternative rhetorical practices and forms of political activity – including activity aimed at radical social and economic change – that are, at the same time, intellectually self-consistent, ethically responsible and practically effective.[30]

Though beset by institutional vulnerabilities and conceptually haunted and hobbled along the lines suggested here, the project of feminist epistemology is not, I think, inherently doomed. A number of recent efforts are especially encouraging, among them an essay by Linda Martin Alcoff titled 'Is the Feminist Critique of Reason Rational?' Arguing against the idea that the missions and practices of philosophy and those of feminism are incompatible, but also against shaky efforts by other feminists to stake their political claims on standard philosophical grounds, Alcoff calls for an alternative, explicitly critical, relation between feminism and philosophy, including epistemology:

> If we [feminists] . . . acknowledge that forms of rationality . . . are embedded within history, we must also acknowledge that reasoned argument is only a part of what is contained in our or any other philosophical writings . . . [T]he better alternative is to reconfigure the relationship between . . . reason and its others, to acknowledge the instability of these categories and the permeability of their borders, and to develop a reconstructed notion of reason . . . as including multiple forms and operating on many levels.[31]

Such a project, Alcoff observes, would be

> incorrectly interpreted as a reduction of reason to unreason . . . Rationality does not need the Manichean epistemic ontology of an absolute truth-mastery over an abject unreason. It needs distinctions, between true and false, and more and less rational, but these can be formulated differently through developing an account of the situatedness of truth and reason.[32]

These (despite some arguably gratuitous retentions) are, I would agree, among the more promising lines to be pursued: distinctions acknowledged but formulated differently; accounts developed, but by appropriating and extending, not blunting, the intellectually radical force of other contemporary critiques and alternatives, including the idea – crucial to but crucially hedged by Harding and Haraway – of the 'situatedness' (that is, historical and contextual contingency and specificity) of what are, at any time, called truth, reason, facts or knowledge.

Where radically new ideas and strong critiques that are seen as intellectually compelling (for example, those of constructivist epistemology and contemporary sociology of science) are also seen as conflicting with accepted (for example, progressive) political theory and practice, the proper response, I think, is not to renounce those ideas, muffle those critiques, or, in order to safeguard politics-as-usual, strive to confine their reach. On the contrary, the most intellectually responsible response and also the one most likely to be, in the long run, politically desirable is, I think, to pursue the implications of those new ideas and critiques as rigorously and extensively as possible, including into the domains of politics – both theory and practice – themselves. The risk, of course, is that the resulting political theory and practice may look quite different from their former as-usual versions: they may, for example, evoke or involve new sets of considerations, different configurations of interests, and different forms of activity for achieving whatever goals are seen as significant. (It is also likely that articulations of those new ideas and related practices – for example, those of science studies – will themselves be transformed, perhaps radically so, in the process.) Such risks for habitual ways of thinking and of practising politics, including radical politics, could also be seen, however, as offering the possibility of substantial – perhaps revolutionary – benefits.

IV

As was mentioned above, Code's understandings of the political implications of constructivist views are by no means unique. Virtually the same set of claims, charges and rhetorical gestures can be found not only in the writing of other feminist epistemologists but also in that of other contemporary philosophers more generally, where, however, they commonly serve other or additional agendas. Thus in a recent book provocatively titled *The Disorder of Things: Metaphysical Foundations of the Disunity of Science*, philosopher of science John Dupré, declaring intellectual kinship with Wittgenstein and Feyerabend and deriving authority from their counter-establishment epistemologies, nevertheless assails 'the sociology of knowledge movement' as follows:

> By asserting that all scientific belief should be explained in terms of the goals, interests, and prejudices of the scientist, and denying any role whatsoever for the recalcitrance of nature, it leaves no space for the criticism of specific

> scientific beliefs on the ground that they do reflect such prejudices [racism, sexism, and so on] rather than being plausibly grounded in fact. The uncongeniality of the sociology of science program to thinkers genuinely concerned with political influences on scientific belief is nicely stated by the feminist philosopher of science Alison Wylie . . . 'Only the most powerful, the most successful in achieving control over the world, could imagine that the world can be constructed as they choose.'[33]

This passage is of interest not only because it recalls the quite similarly worded one by Code examined above (and repeats comparable mischaracterisations of contemporary sociology of science, which certainly does not '[assert] that all scientific belief should be explained in terms of the goals, interests, and prejudices of the scientist') but also because it exemplifies a more general strategy that has become fairly common in current academic controversy: that is, the validation of intellectual traditionalism by appeals to or gestures of solidarity with political radicalism.[34] Notable in that connection is Dupré's curious allusion to 'the uncongeniality of the sociology of science program to thinkers genuinely concerned with political influences on scientific belief' – curious because the work of that program's most eminent practitioners (for example, David Bloor, Barry Barnes, Andrew Pickering or Steven Woolgar) would certainly seem to be extensively concerned with 'political influences on scientific belief'. (Indeed, in the view of various recent detractors, it is quite menacingly concerned with nothing else.[35]) Has Dupré failed to notice that concern? Or is he suggesting that the sociologists are just faking it (not '*genuinely* concerned')? Or is it not, rather, that the particular ways in which such sociologists trace and articulate the complex, dynamic relationships among individual, social and political interests, institutional configurations, technical practices, and scientific statements are hard to square with the traditional claim of philosophy of science to distinguish, on strictly rational or logical grounds, between the truly epistemic and the merely social or political? If so, then the issue here is not the political authenticity of the sociology of science but, rather, the ability of traditional philosophy of science to meet the intellectual challenges of an alternative and to some extent rival project.[36]

A word more may be added on the quoted passage – which should go without saying but evidently does not. Contrary to the charge that Dupré lodges here and finds 'nicely stated' by feminist Wylie, sociologists of knowledge do not characteristically imagine that the world can be constructed as they or anyone else wishes. For, of course, what they characteristically do *as* sociologists of science is investigate how beliefs about the world are socially shaped, constrained and stabilised. Nor do they claim or proceed as if they believed that all things are infinitely malleable by the human mind. The idea that the 'constructed' in 'constructivism' means made-up-in-your-individual-head-however-you-want-it-to-be is a rather vulgar error, to be expected, perhaps, from journalists or from academics in remote disciplines but not, generally, in the work of presumably knowledgeable philosophers of science. As for the idea – floated by Wylie, endorsed by Dupré, and echoed in Code – that constructivist-as-relativist sociologists of knowledge can think the way they do only because they (versus women? versus feminists? versus realist philosophers?) are 'supremely powerful and privileged', 'the most powerful, the most successful in achieving control over the world': well, when one recalls the perennially underfunded, administratively threatened and institutionally precarious situations of various associate or even full professors of sociology of science at places like Urbana, Illinois or Loughborough, England, it appears pretty ludicrous. Of course, as Western, male (where they are) academics, such sociologists may be *relatively* powerful and privileged. But '*supremely* powerful and privileged', 'the most successful in achieving *control over the world*', and so forth? What is the point of such language?

V

I turn now to some final observations. Equivocation, rightward-veering middle-way steering and ritual exorcism are conceptual/rhetorical practices that evade cognitive stress and professional peril but entail their own risks and costs. My main interest in examining such practices here has been to delineate the microdynamics of certain features of contemporary intellectual life. I have also been concerned, however, with the broader implications of those practices, that is, with their communal and institutional as well as individual risks and costs. I shall, in the remarks that follow, be further concerned with both levels of accounting.

Under conditions of acute and widespread conceptual clash within some more or less established institutional-intellectual domain, there will always be those – scholars, scientists, theorists and so forth – whose conviction of the adequacy of traditional views remains unshaken and who, accordingly, will continue to reaffirm the traditional positions as such, staunchly rehearsing the classic justifications and refutations, unruffled by what colleagues experience as the logical bite of current challenges, unimpressed by what others see as the revolutionary implications of those challenges for ongoing practices in the field. I discuss the psychological dynamics of such reaffirmations elsewhere under the term *cognitive self-stabilisation*, and point out there the forms of logical and conceptual strain that often attend them: in brief, a tendency toward continuous – though often artful and rhetorically as well as cognitively effective – circularity.[37] This is not to suggest, however, that a resistance to novelty and adherence to classic but now questioned ideas reflect an intellectual pathology. On the contrary, they may, under some conditions, reflect due confidence and intellectual perspicacity of a high order. For of course, one could hardly maintain *generally* – in regard to ideas any more than to other human products or practices – that the new will always emerge as better; the questioned, as vulnerable; the traditional, as properly superseded.

In any case, no less interesting for a general account of intellectual dynamics are the responses of scholars, scientists and theorists less firmly wedded to the orthodoxy under siege but perplexed in other ways: those, for example, who are unsettled by the current challenges, or at least notice that the classic justifications and refutations are no longer adequate or conclusive for significant portions of the relevant communities, but who, for often complex reasons, are nevertheless not prepared to pursue explicitly – much less endorse wholeheartedly – the controversial new alternatives. Among this latter group are those who, by reason of personal history or current commitments, are especially anxious about the unhappy ethical or political consequences said to be entailed by those alternatives (quietism, complicity, 'anything-goes'-ism, and so forth) or who, as members of a perennially intellectually suspect group (for example, blacks or women), are especially anxious about their own intellectual standing among professional associates. The result in such cases may be, and often is, a compulsive generation of equivocal positions – positions that signal simultaneously or in swift oscillation both an appreciation of and a distancing from currently heterodox ideas. Equivocation under such circumstances may be required for minimal cognitive well-being and immediate professional self-preservation. Since, however, the hybrid positions so generated tend to be conceptually unstable, requiring continuous shoring up and other repair, they are also intellectually costly. Moreover, because prevailing views in a field are always ultimately responsive to broader intellectual developments, the continuous reaffirmation of conventionally accepted ideas, even if only in equivocal terms, risks not only intellectual confinement but professional immobility as well. In other words, what is calculated as prudent equivocation may be, in the long run, professionally as well as intellectually self-disabling.

When radical alternatives to orthodox views are being proposed, it can be strategically useful, though in different ways, for proponents of *either* side – that is, traditionalists *or* revisionists – to cast their own positions as a middle way between extremes. For the traditionalist, it permits a display of genial sophistication and avoids the stigma of sheer stuffiness. For revisionists, the claim of middle-way moderation may dampen automatic alarm signals and retain the attention of audiences who might otherwise just baulk or bolt. Thus a novel and ultimately radical set of ideas may sweeten its debut statement and disarm criticism along certain predictable lines by declaring itself a middle way between the orthodoxy that it does indeed reject (for example, classical realism or normative epistemology) and the heresy, real or phantom, with which it can expect to be identified (for example, classical idealism or everything-equally-good relativism).[38]

Not all audiences, however, will find the middle-way steering either appealing or admirable. The risk for the equivocating traditionalist is that his or her genial sophistication will be seen by more stalwart colleagues as contamination by – or outright capitulation to – mere intellectual fashion. Conversely, the moderating manoeuvres of the prudent revisionist are likely to be regarded by more explicitly radical colleagues as granting too much to established views or as simply regressive. There are, moreover, other – perhaps less visible or audible – members of the intellectual community who will also be impatient with these cautiously equivocated articulations. Especially significant among these are younger practitioners in the fields in question, for whom the most eagerly sought reports of new methods or models will be those that stress, isolate and elaborate their most innovative elements and, thereby, make them most readily accessible for exploration in connection with those practitioners' own, perhaps comparably innovative, projects. To put this in broader terms: the versions of new intellectual developments that indicate most clearly and powerfully their most heterodox elements and challenging implications are also, for that very reason, the ones most likely to energise investigation, extension and refinement – and, by the same token, to reveal most readily their conceptual and practical limits or inadequacies.

In relation to such broader perspectives, a number of the moves and strategies discussed above, including devil-blackening, political flag-waving and tendentious mischaracterisation, could be seen as not merely individually but communally – socially and institutionally – costly as well. Rhetorical strategies are not, in my view, contemptible as such, or necessarily otherwise objectionable. On the contrary, I cannot imagine what a non-strategic argument or non-rhetorical exposition would be. Nor do I think conceptual moves can or should be 'free', as it is said, of institutional, professional or even social pressures. Indeed, I would say that a susceptibility to being shaped by such pressures is inseparable from the virtue of intellectual responsiveness or, perhaps, just another way of stating that virtue. I do not, then, set conceptual moves and rhetorical strategies in opposition to right-thinking and straight-talking. Nevertheless, I do think the intellectual life of a community can become limited, maimed and stultified when easy points are scored by appeals to ignorance; when concepts, positions, individual figures and entire disciplines are – without having been intellectually engaged – casually maligned or swaggeringly derided; and when political or moral pieties are used in cloak-and-dagger operations against intellectual or institutional rivals.

Finally, as I hope is clear, I am not calling here for an across-the-board show of intellectual heroism. Different situations require different assessments of personal, professional and social risk, and, of course, some vulnerabilities are more visible than others. My broader purpose here has not been to indict individual scholars or theorists for trimming or timidity but, rather, to highlight some of the conceptual and rhetorical practices that both reveal and perpetuate the risks of intellectual radicalism in our own time and to suggest some of the costs we all pay as a result.

Notes

1. Putnam, quoted in an advertisement for Susan Haack, *Evidence and Inquiry: Towards Reconstruction in Epistemology* (London: Blackwell, 1995) in *The Times Literary Supplement*, no. 4843 (26 January 1996), p. 11.
2. Riordan, 'Trading Information', p. 38.
3. B. C. Smith, *On the Origin of Objects*, pp. 3–4.
4. Smith cites Bruno Latour's 'Irreductions' (see Latour, *The Pasteurization of France*, pp. 153–236).
5. B. C. Smith, *On the Origin of Objects*, pp. 111–12, italics in text.
6. Ibid., p. 195.
7. Haraway, *Simians, Cyborgs and Women*, p. 187. Cited and quoted in B.C. Smith, *On the Origin of Objects*, p. 4n2.
8. For examples of such developments, see Biagioli (ed.), *The Science Studies Reader*. For other examples and discussion, see B. H. Smith and Plotnitsky (eds), *Mathematics, Science, and Postclassical Theory*.
9. Thus Smith's *On the Origin of Objects* is described on the book's jacket as 'a sustained critique of the formal tradition underlying reigning views' of cognition. Similarly, Andy Clark's *Being There: Putting Brain, Body and World Together Again*, discussed below, is said by its publishers to introduce a 'paradigm shift' in our understanding of the mind.
10. Clark, *Being There*, pp. 220–1.
11. Clark speaks in the same passage of the need to 'anchor', in knowledge of the brain, a 'canny combination' of ecological approaches and traditional cognitive science (*Being There*, p. 221). It is not, I should stress, the stylistic lapses as such that concern me here but the conceptual problems that they seem to signal.
12. Clark, 'Time and Mind'. As in *Being There*, he cites Thelen and L. Smith, *A Dynamic Systems Approach to the Development of Cognition and Action*; Hutchins, *Cognition in the Wild*; and, especially significantly here,

van Gelder, 'It's About Time: An Overview of the Dynamical Approach to Cognition'.

13. For an instructive comparison of major alternative contemporary cognitive theories by a proponent of dynamical/ecological models, see Hendriks-Jansen, *Catching Ourselves in the Act.*
14. B. C. Smith, p. 97.
15. See Chapter 3 above.
16. For an illuminating discussion, see Oyama, *The Ontogeny of Information*, pp. 5, 33–4.
17. Rorty, *Objectivity, Relativism, and Truth*, pp. 21–34.
18. Latour, *We Have Never Been Modern*, pp. 61–2.
19. Damasio, *Descartes' Error*, p. 280.
20. Harding, 'Rethinking Standpoint Epistemology: What is "Strong Objectivity"?', p. 50. Harding argues that what would be more 'strongly' objective than knowledge claiming to be objective in the conventional 'weak' sense (that is, unbiased by personal or political interests) was knowledge produced (or ratified) by democratic procedures that secured the input of women and, in principle, members of all other politically marginalised populations. Significantly here, her articulations of the argument (strained, in my view, from the outset) tend to be severely hedged and, on crucial points, vague. Indeed, as Harding explicates and qualifies her views and proposals in successive versions, they appear increasingly indistinguishable from conventional objectivist/realist philosophy of science except for their association with an explicitly progressive (or at least *bien-pensant*) political perspective and agenda, but no less dubious or elusive than before in practical terms. For recent versions, see Harding, *Is Science Multicultural? Postcolonialisms, Feminisms and Epistemologies.*
21. Lorraine Code, 'Taking Subjectivity into Account', p. 39.
22. Ibid., p. 39.
23. For the force of the parenthetical presumptions here, see the discussion of relativism in Chapter 3, Section IX. It is only when validity is understood along classical lines as an objective, noncontingent attribute that the notion of the equal validity of all beliefs appears absurd: that is, as implying that all beliefs are always equally *objectively* true. Otherwise the notion could be seen as a tautology or truism – that is, as saying in effect either that all beliefs are really believed by those who believe them or that everything that has been believed by some people under some conditions has had some kind of cogency, legitimacy, effectiveness etc., *for those people under those conditions*. These observations apply as well to the other egalitarian claims ('all equally good, equally beautiful' etc.) supposedly implied by critiques of classical objectivist views. I discuss this strictly non-sequitur reasoning (that is, the derivation of a claim of *objective equality* from a critique and rejection of *objectivism*) elsewhere as the Egalitarian Fallacy. See B. H. Smith, *Contingencies of Value*, pp. 98, 152, 157 and *Belief and Resistance*, pp. 77–8.
24. For recent examples of such practices in connection with the 'science wars', see Gross and Levitt, *Higher Superstition*; Sokal and Bricmont, *Fashionable Nonsense*; Koertge, 'Scrutinizing Science Studies'; and Boghossian, 'What the Sokal Hoax Ought to Teach Us'. They are discussed in Chapter 5 below.

For a good sample of the array of tactics, see Boghossian, 'Constructivist and Relativist Conceptions of Knowledge in Contemporary (Anti-) Epistemology: A Reply to Barbara Herrnstein Smith', a commentary on an earlier version of the present chapter. I reply in B. H. Smith, 'Reply to an Analytic Philosopher'.

25. Code, 'Taking Subjectivity into Account', p. 42.
26. Ibid., p. 40.
27. Bloor, *Knowledge and Social Imagery*, p. 7. See also Note 23 above.
28. See, for example, Cussins, 'Ontological Choreography: Agency through Objectification in Infertility Clinics'.
29. See the discussion of Sandra Harding above, and Note 20.
30. See B. H. Smith, *Belief and Resistance*, pp. 10–15, 26–35.
31. Alcoff, 'Is the Feminist Critique of Reason Rational?', p. 75.
32. Ibid.
33. Dupré, *The Disorder of Things*, pp. 12–13. He cites Alison Wylie, 'The Interplay of Evidential Constraints and Political Interests: Recent Archaeological Research on Gender', pp. 15–35.
34. Dupré's effort throughout the book is to establish his own middle-way position of 'moderate realism' as anti-logical-positivist but realist, rationalist and committed to the normative projects of traditional philosophy. A philosopher of science himself, he appears especially concerned with current threats to his field so defined (that is, as realist, rationalist and normative), both from 'trendy' scientistic, as he sees them, projects within philosophy (for example, the effort to naturalise epistemology) and also from apparently encroaching moves in other disciplines, for instance, cognitive science, neurophysiology, sociobiology, and, as here, sociology of science (see *Disorder of Things*, pp. 13 and 268n12).
35. See, for example, Gross and Levitt, *Higher Superstition*, pp. 42–70.
36. Dupré's vague wave of the hand, immediately after, toward an 'interaction' between 'the role of social, political and personal factors' and 'some account of justifiable belief', simply reasserts the standard normative/positivist definition of knowledge at issue (that is, as justified/able belief), as does also his suggestion that 'the only way to put the genuine insights of the sociology of knowledge program . . . into a proper perspective is to see the kinds of forces they describe as interacting with a real and sometimes recalcitrant world' (Dupré, *Disorder of Things*, p. 13). Though evidently intended to suggest a moderate best-of-both stance, these gestures, in assuming precisely the dichotomy that sociologists of knowledge reject – that is, between (mere) 'social forces' and a presumably autonomously corrective ('sometimes recalcitrant') 'real world' – are effectively empty.
37. See B. H. Smith, *Belief and Resistance*, pp. 88–124.
38. See, for example, Maturana and Varela, *The Tree of Knowledge: The Biological Roots of Human Understanding*, pp. 133–5, and (as discussed above but considered here as more fundamentally radical and determinedly strategic) B. C. Smith, *On the Origin of Objects*, pp. 3–4.

Chapter 5

Disciplinary Cultures and Tribal Warfare: The Sciences and the Humanities Today

The durability of C. P. Snow's notion of 'the two cultures' is a testament, no doubt, to the evocativeness and apparent continued aptness of the phrase, but also, one suspects, to the sense of scandal that has always attended it: its acknowledgement, that is, of extensive ignorance and provincialism among the educated classes and its image of the academy as divided into two mutually suspicious or indeed warring tribes. The intellectual map has shifted in important ways since Snow's essay was first published, and his account of the differences between natural scientists and the group he called 'literary intellectuals' appears increasingly dated and, indeed, itself quite provincial.[1] One notes, for example, the academic and intellectual significance of the social sciences, which Snow famously overlooked; the emergence of such fields, disciplines and interdisciplines as women's studies, environmental studies, media studies or cognitive science; the increasing intellectual and social importance of the biological vis-à-vis the physical sciences; and the increasingly close collaboration between academic research institutions and industrial and governmental units.[2] Transformations of these kinds over the past fifty years, along with other institutional and intellectual developments discussed elsewhere in this book, have significantly complicated the individual situations of and mutual relations between the natural sciences and the humanities – neither of which was ever simple, stable or uniform – and also their respective characterisations, both as assumed and articulated by practitioners and as formally theorised in such fields as epistemology, education theory and philosophy of science. Nevertheless, elements of a two-culture divide – certainly the *ideology* of such a divide – remain with us. They figure conspicuously, for example, in the series of provocations and skirmishes now commonly referred to as the 'science wars' and more obliquely but still powerfully in the practices and self-promotions of the currently high-profile field of evolutionary psychology. Both of these sites of a two-culture residue will concern us in this chapter. The latter will be explored further in Chapters 6 and 7.

With regard to the 'science wars', my purpose here is not to make a case for one or another of the supposed 'sides' in the multiple and often quite varied conflicts thus named. I do consider, later in this chapter, the substance and justice of some of the sensational charges and alleged exposures involved. One of my major points, however, is that the contending parties are not 'humanists' as such versus 'scientists' as such.[3] Nor are they, I think, the forces of enlightenment versus those of darkness and superstition; nor the sensible, responsible people versus the crazies, frauds, and ideologues; nor, for that matter, the forces of culture, sensibility and virtue versus those of cold, amoral technocracy. There are certainly important conflicts being played out in the contemporary intellectual-academic world. As is commonly the case in all battles, however, the *naming* of the conflict is itself a central issue in it, as is the representation of the parties involved and of their respective positions. What I hope to do here, rather, from my perspective as a long-time observer of intellectual controversy more generally,[4] is to offer some observations on the current debates that may help illuminate the more aggressive energies sometimes exhibited in them. At the end, I shall also suggest why, despite the significant divergences and strong antagonisms involved here, current proposals to 'integrate' the disciplines offer neither a historically plausible project nor an intellectually or otherwise desirable ideal.

I

The chart below, a schematic representation of 'the two cultures' in more or less classic form, serves as an entry point for exploring these issues. It is intended to suggest how scientists and humanities scholars, at least a good many of them, might characterise their own aims, methods, values and achievements and how, at least when they were feeling genial and generous, they might see their respective differences from one another. This is not, it should be noted, the picture presented by Snow himself, who was more interested in distinguishing character types ('scientists' and 'literary intellectuals') than types of academic culture and whose way of drawing the differences between them was by no means even-handed (as this chart seeks to be).[5] The terms 'etic' and 'emic' that head the two columns are borrowed from cultural anthropology, where they distinguish between the kind of description, explanation or interpretation of some cultural practice (say, an initiation ritual) that might be given by a visiting anthropologist and the kind that might be given by a native member of the culture in question.[6] The terms are handy here because, in the idea of an outside ('etic') impartial observer as distinguished from an inside ('emic') sympathetic participant, they sum up a recognisable conception of the relation between the sciences and the humanities – that is, the conception, first proposed in the late nineteenth century by a group of German philosophers, of the relation between *Naturwissenschaften* and *Geisteswissenschaften* or, as we could translate the paired terms here, between the disciplined study of *natural* phenomena and the disciplined study of *cultural* phenomena.[7]

A Schematic Representation of 'The Two Cultures'

Etic	*Emic*
Observer's perspective	Participant's perspective
Naturwissenschaften, natural sciences	*Geisteswissenschaften*, humanities
Study of non-purposive, material objects and events, natural phenomena	Study of purposive, meaningful actions and creations, texts and artifacts
Careful observation, precise measurement and calculation	Careful examination, close reading, rigorous reflection
Classification, description, explanation	Interpretation, evaluation, explication
Descriptive, value-neutral	Normative, critical
Data collection, experimentation, construction and testing of hypotheses	Archival research, textual and conceptual analysis, development of interpretations
Seeks knowledge, prediction, control	Seeks understanding, insight, appreciation
Yields models, causal accounts, general theories	Yields analyses, commentaries, critiques, reconstructions
Indicates regularities, facts, laws	Indicates connections, values, possibilities
Stresses simplicity, clarity, quantification	Stresses richness, depth, subtlety
Replicable by other scientists	Appropriable by other scholars
Intellectually valuable, technologically advantageous	Intellectually valuable, culturally enriching

As one glances down each of the two columns, most of the items and classifications should appear plausible and mutually coherent: on the one side, the association of the natural sciences with observation, description, models, general laws, quantification, replication and technological application; on the other side, the association of the humanities with close reading, critical commentary, interpretation, appreciation, insight and so forth. Also, as one looks across the columns at the corresponding entries for each side, they should appear appropriately matched and reciprocal – that is, as reflecting differences and perhaps tradeoffs but not, or not necessarily, conflicts or rivalries. Thus *knowledge* in the natural sciences matches *understanding* in the humanities; the *simplicity* and *clarity* valued in scientific pursuits complement the *richness* and *subtlety* valued by humanists; and so forth. Presumably these are, on each side, good things: values and virtues, worthy aims and proper methods. Moreover, it could be said that these two sets of enterprises, depicted in such terms, reflect, respectively, two distinct but equally fundamental ways in which, as human beings, we relate generally to the world around us: that is, as inevitably *both* observers (including self-observers) and participants, explainers and interpreters; as creatures who seek to extend our knowledge of and control over the material conditions of our existence but who seek also to develop and enlarge ourselves as experiencing, reflective beings and who, as such, are engaged by the experiences and reflections of our fellow beings.

It's a nice picture, I think, and by no means grossly false, but perhaps too neat or, as some readers may feel, *too* evenhanded or symmetrical. Indeed, if we linger a bit over some of these items, we may begin to note, first, that the intellectual world in which we operate is rather more complicated than what is depicted here and, second, that there is a certain degree of instability in the paired items, such that the differences do not remain nicely balanced or neutral but become more or less clearly valenced, implicitly hierarchical, and potentially invidious. For one thing, many individual scientists and humanists are likely to object – quite properly, in my view – that their own aims and achievements are by no means restricted to the columns to which they are assigned here but share many or most of the attributes listed on the other side. After all, it might be asked, doesn't science produce insight and understanding as well as models and technologies, and don't scientists, too, value imagination and respect the complexity of phenomena? Correspondingly, aren't clarity and rigour valued as much among responsible scholars as among responsible scientists, and don't researchers in fields such as art history or literary studies produce new knowledge? Also, someone is sure to notice – again, quite properly in my view – that the *social* sciences seem to have been left out in the cold here and that the very existence of fields such as anthropology, economics, or sociology undermines the idea of just *two* disciplinary clusters or types of epistemic culture – or, as in Snow's version, two types of intellectual. And, perhaps most significantly, sooner or later we will recall the many phenomena that are not clearly 'natural' as distinct from 'cultural', or 'human' as distinct from 'natural': *language*, for example, which is certainly human (even if not exclusively so) and both 'natural' and 'cultural'; or *culture* itself, which could certainly be seen as a naturally occurring phenomenon (among human beings, along with some other species) and thus properly the object of study by natural and/or social scientists (the latter thereby receiving explicit acknowledgement); or, indeed, *science* itself, which could certainly be seen as, among other things, a set of historical phenomena, cultural practices and human creations and thus properly the object of study by both social scientists and humanists.

Indeed, it begins to appear that virtually any of the academic disciplines could lay claim, in one or another of its aspects or pursuits, to virtually all the aims, methods, achievements and domains of study that are here divided and contrasted. Before settling too quickly, however, into that alternative image of happy heterogeneity (or, as some might see it, intolerable chaos), we must acknowledge the recurrent duplication *within* many of the disciplines of just the sorts of conceivably complementary but also potentially antagonistic divisions represented on our chart: the division, for example, between 'biological' and 'cultural' anthropology, or between 'experimental' and 'clinical' psychology, or, in literary studies, between, on the one hand, historical or philological 'scholarship' and, on the other, interpretive 'criticism' or 'theory'. As most academics discover quite early in their careers, people who follow different approaches in nominally the same discipline often find one another's activities and achievements as alien, idle or otherwise objectionable as some scientists find the activities and achievements of all humanists – and vice versa.

II

This continuous generation of disciplinary dualisms and antagonisms is not, I think, the sign of some underlying fissure in the universe or the human psyche. It does suggest, however, the operation of strong and recurrent forces – forces, I think, that reflect quite general psychological tendencies and social dynamics but that are also fed and shaped by specific historical, cultural and institutional conditions. It seems clear, for example, that the kinds of mutual antipathy displayed in the current 'science wars' and in Snow's not-so-even-handed delineation of 'the two cultures' draw on impulses that are both endemic and, in some respects, extremely primitive. Among them is the identity/difference polarity: that is, our tendency, in identifying ourselves with one or more social group (ethnic, religious, political, professional and so forth), to experience that identity through contradistinction to one or more other groups – or, in other words, to experience the world in terms of 'us' and 'them'.[8] Another such impulse is what I describe elsewhere as the tendency to self-standardising and other-pathologising – that is, to see the practices, preferences and beliefs of one's own group as natural, sensible and mature and to see the divergent practices, preferences and beliefs of members of other groups, especially those one casts as definitively 'other' to one's own, as absurd, perverse, undeveloped or degenerate.[9]

It is relevant here that 'the two cultures' are, in many ways, precisely *cultures*. Quite aside from the specific domains of knowledge or specialised technical practices that distinguish the natural science disciplines from the humanities, there are crucial differences in their prevailing conceptual styles, discursive idioms, measures of success and norms of social interaction. Individual disciplines and fields (microbiology, particle physics, medieval studies and so forth) are, of course, prime examples of what Ludwik Fleck termed 'thought collectives' and, as such, give rise to – and are largely held together by – the systems of shared beliefs, pragmatic routines and perceptual dispositions that he called 'thought styles'.[10] But so also, to some extent, are the larger social-epistemic groups constituted by those who identify themselves as '(natural) scientists' or as 'humanists', with divergent cultural norms ranging from the sartorial (informal versus more or less conservative or chic) to the verbal-stylistic (precision versus resonance). The consequent culture-clashes are familiar to those who frequent interdisciplinary venues. For example, the verbal subtleties and high-culture allusions that may come naturally enough to any duly acculturated humanist are likely to appear wordy, affected or just plain opaque to a natural scientist; conversely, the pared-down functional prose and old schoolroom jokes that win the scientist praise among his fellows as a clear thinker and great wit are likely to strike the literary scholar as dull and juvenile.

Us/them distinction-drawing and related polarised valorisation, which appear to be recurrent features of human sociality,[11] often figure significantly in the processes through which our individual disciplinary identities are formed. The fields of study and practice to which we are drawn commonly reflect our prior (innate and/or already well-formed) tastes, talents and temperaments (for example, a fascination with numbers or pronounced musical or verbal skills) but are also reinforced and sharpened by disciplinary membership as such. Thus the sorts of tastes and talents just mentioned will commonly be encouraged, rewarded and enhanced by explicit training and practice in our field (for example, mathematics, music composition or literary study) and by increasingly extended association with like-minded and like-skilled peers, teachers and mentors. If the personal-professional match was a good one to begin with, we commonly become more of what we initially were, and more thoroughly – and, of course, proudly and perhaps exclusively – so. Given the general tendency to self-standardising noted above, we are likely to regard what we are as a manifestly good thing to be or, indeed, as quite the better, best or in some cases *only* thing to be: a member of the military, not a more or less pathetic civilian; a poet, not a damned shop-keeper; a scientist, not a more or less befuddled nonscientist.

The various polarising tendencies described above draw substance as well as force from other dualisms pervasive in the general culture: for example, 'deeds' over 'words', 'serious work' over 'mere play', 'logic' versus 'emotion', 'hard' versus 'soft', and, we need hardly add here, 'masculine' versus 'feminine' or 'effeminate'.[12] The relation of such loaded binaries to the ideology of the two-culture divide is unmistakable and often unapologetically explicit. (Snow, for example, lauded the tone of the culture of scientists for being 'steadily heterosexual' and, in contrast to the literary culture, free of 'the feline and oblique'.[13]) We begin to see, then, how our designedly even-handed, complementary etic/emic chart might transform itself into an antagonistic, hierarchical and thoroughly invidious one. To a strenuously partisan humanist, for example, it might appear as an apt summary of the differences between, on the one side, the properly honoured qualities of subtlety, depth, responsiveness and imagination found among people in the humanities and, on the other side, a good description of everything alien about those tin-eared, literal-minded people over in science. Or, conversely, a strenuously self-congratulatory scientist, looking at exactly the same array and arrangement, might see it as a good summary of the fundamental differences between, on the one side, the properly honoured scientific virtues of rationality, objectivity and tough-mindedness and, on the other, all that touchy-feely, fuzzy-minded subjectivism you get over there in the humanities. The primitive us/them impulses and crude caricatures evoked here may, of course, be played out relatively harmlessly, as in the mutual teasing of college roommates or old cronies at the faculty club. They are, however, always distorting and, when nourished by feelings (justified or not) of outrage or injured dignity and joined to a sense of moral crusade, can become quite virulent, self-confining and communally stultifying.

Of particular significance here are the self-perpetuating effects of such caricatures, which, casually invoked by teachers, rehearsed and amplified by fellow students, and echoed emptily but resonantly enough in the media, become 'what everybody knows' without anybody having either the motive or the opportunity to discover otherwise. The mutual hostility and ignorance between scientists and humanists noted by Snow fifty years ago may be, in some respects, even sharper and more extensive now than then, and the institutional, especially educational, mechanisms of their perpetuation even more deeply entrenched. There is more involved here, I think, than ever-increasing specialisation and pre-professionalism, though both are certainly implicated. For one thing, to the extent that 'the two cultures' are, as suggested above, precisely *cultures*, significant differences in their respective cognitive styles, discursive idioms and institutional practices make it difficult – at least immediately – for those deeply acculturated in one of them to operate smoothly and effectively in the other. Also, two-culture crossing is discouraged by the effects of institutional inertia, specifically by academic arrangements that reflect not only established curricular and pedagogic traditions but also the power and pervasiveness of just the sorts of stereotypes discussed above. Thus it is widely believed by students in the humanities today that the aims and operations of the natural sciences are intrinsically intellectually alien to them (mechanical, numerical, reductive and so forth) and likely to be personally alienating as well (aggressively patriarchal, homophobic, complicit with imperialism and racism, dehumanising and so forth). At the same time, it is believed by many faculty in the sciences that teaching and scholarship in the humanities consists of idle opinion-mongering, that humanities students – whose heads (I've been assured) are 'accustomed to mush' – are incapable of the 'hard work' and 'rigorous thinking' required to do 'real science', and, consequently, that the only alternative to the sorts of pre-professional science courses those faculty are already teaching would have to be something 'watered-down' accordingly. And, to round out the picture, virtually pre-emptive course requirements for students majoring in the natural sciences leave them little time to explore such fields as history, philosophy or literary study, so that those among them who arrive with interests along such lines commonly cannot pursue them while those who arrive already scornful (like many of their peers and professors) of the humanities – as, typically, 'soft' or 'useless' – are given little reason to think or occasion to learn otherwise. So it goes: hard work and mere play, hard facts and mere words, hard and soft disciplines, hard and soft thinking, hard and soft people. Crude binaries of this kind are what I have been calling the *ideology* of 'the two cultures'. Their self-perpetuating operations are obvious.

III

As mentioned earlier, there are a *number* of broad struggles taking place in the present academic-intellectual arena, and their relations to each other are fairly complex. For example, the 'science wars', though distinct from the earlier (and by no means ended) 'culture wars', are nevertheless fuelled by some of the same intellectual and demographic changes that provoked the latter, and both are related to several other intellectual and institutional conflicts dating from around thirty-five years ago – which is to say, a generation ago. Indeed, as noted in Chapter 2, many of these conflicts resolve readily into generational struggles – not necessarily between older and younger faculty but between those trained in and committed to more traditional approaches and those familiar (sometimes through self-education) with more recent ones and already working with them. We may recall briefly three especially relevant divisions and antagonisms of this kind:

1. The conflict between, on the one hand, philosophers, scientists and other academics who accept and assume the more traditional (logical empiricist, critical rationalist, scientific realist) accounts of science developed by theorists of the Vienna Circle, Karl Popper and other major figures of twentieth-century philosophy of science and, on the other hand, those engaged by and employing the revisionist (constructivist/pragmatist) accounts developed by such historians, philosophers and sociologists of science as Thomas S. Kuhn, Paul Feyerabend and Bruno Latour;
2. The conflict between, on the one hand, those committed to the classic realist/rationalist/referentialist views of truth, knowledge and language developed and defended in mainstream analytic philosophy and, on the other hand, those well schooled in the challenges to those views – and, significantly, working with the alternatives to them – developed by such eminent but not quite canonical Anglo-American philosophers as William James, J. L. Austin and Richard Rorty and by such continental theorists as Michel Foucault and Jacques Derrida; and, finally (though the list could be extended),
3. The division within most humanities fields, especially literary studies and history, between scholars and teachers committed to the methods and approaches developed in the early- to mid-twentieth century (bibliographic scholarship, positivist historiography, formalist New Criticism and so forth) and those whose work reflects more recent developments in their fields, such as new historicism, feminist and postcolonialist criticism, cultural studies or media studies.

Several points may be stressed here, accordingly. First, it is clear, I think, that one cannot map the set of academic conflicts just outlined onto the two-culture divide; the divisions indicated here do not correspond to the differences, such as they are, between scientists and humanists. To be sure, there still exist some self-consciously high-toned, nostalgic 'literary intellectuals' of the kind Snow described in the late 1950s and whom he set in contrast to his solid, sensible, socially progressive 'scientists'. On the current academic battlefields, however, humanists of that kind are likely to be united with their colleagues in the sciences against the greater perceived menace of 'postmodernism' while the embattled scientists are just as nostalgic as their humanist allies. Also, social progressivism in the contemporary university, especially where it concerns itself not, as in Snow's era, with labour unions, poverty or despotism in foreign places but, rather, with matters of race, ethnicity, gender or sexuality quite close to home, is likely to encounter at least as much hostility from tradition-minded science faculty as from tradition-minded faculty in the humanities, united once again but here against the greater perceived menaces of 'academic feminism', 'multiculturalism' and 'political correctness'.

Secondly, of the three sets of academic conflicts described above, only the first has anything to do with science per se and none of them involves an 'attack on science' as distinct from critiques of certain conventional ideas about science. The latter distinction, analogous to the difference between attacking the stars and challenging the validity of astrology, seems obvious enough, but is often missed or obscured. Of course, to people who think of the stars in astrological terms (that is, as influencing or indeed determining everything from individual character to the fate of nations), a challenge to astrology might look like an effort to diminish the stars' power and thus like an attack on the stars themselves. In the same way, to those who conceive of science as a body of duly privileged knowledge and/or a distinctive method for arriving at such knowledge, challenges to that conception evidently look like attacks on science itself. The stronger and more deeply invested one's conviction of the intellectual propriety and moral or political necessity of certain ideas (about, for example, the nature of truth, knowledge or science), the less able one will be to entertain even the possibility of a genuine challenge to those ideas. Thus carefully argued critiques will look like irrational attacks; broad social concerns will be interpreted as expressions of personal resentment; influential scholars developing those critiques or voicing those concerns will appear to be incompetent or impostors; and students or colleagues citing the work of those scholars will look like dupes of fashion.

Clearly there can be no genuine intellectual engagements where convictions and perceptions of these kinds prevail. Debates over the conceptual, methodological, institutional and broader social or political implications of constructivist views of science are certainly being conducted these days among scholars and theorists in the various fields involved – the history and sociology of scientific knowledge, epistemology, the philosophy of science and so forth; their arguments, analyses and exchanges appear in the pages of such journals as *Social Studies of Science* or *Philosophy of Science*. Debates of that kind, however, are not what make up the 'science wars', which are fought largely by people outside the fields involved and primarily in the pages of newspapers, general interest periodicals and books addressed to non-specialist readers. None of the academics involved in the initial volley of charges in these 'wars' was a historian, sociologist or philosopher of science; and, of course, being a scientist does not make one an expert in science studies any more than being a member of some culture makes one a cultural anthropologist.[14]

This last analogy points to an interesting and perhaps ironic reversal that brings us back to our emic/etic chart. Clearly, one of the reasons that some scientists are disturbed by recent work in science studies is that their practices are treated there from an 'etic' (outsider, impartial) perspective rather than, as in more traditional historical or philosophical accounts, from an 'emic' (insider, sympathetic or indeed reverential) perspective. Scientists, it appears, prefer their own activities and productions (laboratory practices, journal articles, technical inventions and so forth) to be studied *humanistically*, that is, as the products of individuated human subjects whose ideas, intentions and achievements can be duly reconstructed, understood and appreciated by their fellow humans, rather than *scientifically*, that is, as natural phenomena to be observed dispassionately and explained naturalistically – for example, as exemplifying the more general operations of human cognition or as displaying the more general dynamics of social interaction. The antagonism toward recent sociology of science on the part of some scientists can be understood accordingly, as can also, perhaps, the outrage it fuels where it involves a sense of injured dignity.[15] The question at issue, however, is whether such reactions on the part of some scientists (in effect, the 'natives') invalidate the observations and analyses of contemporary science studies (in effect, the 'anthropology' of Western science).

A third point to be stressed here, especially in view of the confusions and determined conflations that abound, is the difference between the specifically *academic* antagonisms outlined above and the sorts of clashes exemplified by the resistance of Christian fundamentalists to evolutionary theory, or the campaign by animal-rights activists against the use of animals in laboratory research, or the scepticism toward NASA voiced by believers in alien abduction. It is hard to see what is gained by lumping all these together with poststructuralist language theory, feminist epistemology and science studies as evidence of a widespread current 'attack on science' except nourishment for feelings of paranoia among scientists – or, to be sure, a large sale of books to readers who are just educated and enlightened enough to enjoy feeling contempt for the uneducated and the zealous, are not too finicky about sources and citations, and are happy to have a chance to laugh at such naturally ridiculous people as literary theorists, feminists and French philosophers. Just why these last are *naturally* ridiculous returns us to the more primitive impulses mentioned earlier, which often display themselves – among the officially educated and enlightened as well as elsewhere – as anti-intellectualism, misogyny and xenophobia.

IV

It may be asked, of course: is there *no* justice to any of these charges, no grounds for outrage on the part of people like Sokal or Gross and Levitt? The answer, I would say, is yes, *some* justice, but not much, and no, *no* grounds for a show of ostensibly intellectual outrage that has, as I have been suggesting, other sources. Yes, there are scholars in the humanities (today as ever) whose knowledge of science is quite limited, and some of them do make inane, politically tendentious or otherwise dubious statements about science. There are also philosophers, critics and writers – both inside and outside the academy – whose knowledge of science may be relatively extensive but who occasionally make technically inapt, metaphoric or simply nonstandard use of concepts developed in science.[16] It is rather a stretch, however, to represent the latter as engaged in an 'abuse' of science or to take the work of any of these scholars or writers as evidence of an anti-science crusade on the part of 'the academic left', 'postmodern French intellectuals' or 'humanists' generally. As for the scandalising sound bites routinely associated with such groups or attributed to the thinkers identified with them (for example, 'truth is relative', 'reality is a social construction', 'the author is dead', or 'anything goes'), one finds, on inspection, that it is not what they say at all; or that they say it but with a meaning quite different from that inferred or imputed; or that they say it but its absurdity – like 'the earth goes around the sun' or 'people are descended from apes' – consists simply and entirely of its being contrary to currently received wisdom.

What, then, remains of these charges? Very little, it appears. Indeed, the 'science wars' can be seen as something of a mirage: arising from ignorance and arrogance; amplified by opportunism, both academic and journalistic; and fought against a largely phantom enemy with much artillery fire but few strikes. The sensational charges were clearly scattershot and often, on inspection, empty. The supposed exposures consisted, on inspection, of misleadingly framed pastiches or garbled accounts that exposed little in particular and nothing in general, certainly nothing about 'intellectual standards in the humanities' or the value, intellectual or other, of any of the ideas, movements or approaches ostensibly targeted.[17] Another question may be relevantly posed here, however: namely, why did those largely ill-aimed and often empty charges gain such wide and rapid credibility among members of the educated public?

One reason, clearly, is that the process of weighing the merit of charges and counter-charges in these battles was complicated by the cross-disciplinary character of the battles themselves. Ordinary criteria of intellectual assessment and control that operate routinely and relatively effectively *within* a field (one thinks, for example, of the letters-to-the-editor columns in a professional journal such as *Science* or *PMLA*) tend to break down or function erratically when claims and indictments concern matters outside the speaker's own field, especially when laid before an audience of non-specialists. Indeed, because the grounds, signs and measures of intellectual authority were themselves key issues in these controversies (who were the 'experts' and who the 'ideologues'? who was/is authorised to speak *about* science at all?), exaggerations, distortions and misrepresentations were not caught and de-activated as they might have otherwise have been but continued to operate as live ammunition – that is, as supposedly authoritative reports or exposés, available for recurrent citation, invocation and amplification – in the intellectually void no-man's-land between the warring cultures.[18]

A second related reason for the lopsided outcome of the 'science wars' returns us to the two-culture stereotypes discussed above. Where the weight of intellectual credentials is in doubt, the power of prevailing biases is likely to be all the stronger. It may be true, as scientists sometimes complain, that the image of them presented in the popular media is often unflattering. (Thus the author of an article in a recent issue of *Science* writes, in some distress, that science-fiction movies and children's television give the public the 'tacit message' that 'scientists are . . . heartless, humorless nerds').[19] Nevertheless, the intellectual authority of the natural scientist is rarely in doubt for a general public audience. 'Einstein', 'brain surgeon' and 'rocket scientist' remain vernacular reference points for extremes of intellectual prowess and technical expertise, and a Nobel prize in physics is pretty solid intellectual capital in most venues. My point here is not that scientists enjoy an especially good press. It is, I think, actually a quite mixed press – the same as for politicians and English professors. My point, rather, is that, when authorities clash across the two-culture divide, the benefit of the doubt is likely to be given to the scientist. We might note here the evidently general assumption that, when a humanist (say, a professor of classics) fails to understand some scientific matter (say, the Second Law of Thermodynamics, as in Snow's notorious example), then it must be the humanist's fault (deplorable ignorance, indifference, ineptness and so forth), but when a scientist fails to understand the work of a humanist (say, the writings of some French philosopher), then it must *also* be the humanist's fault (deliberate obfuscation, intellectual fraudulence and so forth). In literary studies or philosophy, however, just as in physics or mathematics, the ability to appreciate the force of new ideas – or, for that matter, old ideas – requires a more than casual understanding of the issues involved, a more than casual awareness of the history and contemporary significance of alternative positions, and an active, current mastery of the relevant technical skills, for example, mathematical or textual. A certifiably high IQ plus a love of Shakespeare do not qualify one to appreciate Derrida's reflections on Heidegger any more than a certifiably high IQ and the ability to balance one's bank accounts qualify one to appreciate the proof of Fermat's Last Theorem.

Most humanities scholars, like most scientists, assume a certain degree of relevant sophistication in the audiences they typically address – that is, other scholars in their fields. Accordingly, unless they are writing specifically for a general audience, humanists (literary scholars, philosophers and so forth) as well as scientists (particle physicists, organic chemists and so forth) assume a more or less knowledgeable and appropriately trained readership. Just as humanists are not commonly familiar with the specific technical concepts or mathematical representations used by scientists addressing other scientists in their fields, scientists do not commonly have the ranges of specific historical knowledge or types of specific textual skills that scholars in the humanities can generally count on in readers in *their* own fields. The parallels and symmetries here would be obvious enough to anyone even minimally familiar with the array of academic disciplines, for example, an intellectual historian or (one hopes) university upper administrator. They are far from obvious, however, to a majority of the general public or, it appears, to a good number of scientists, both of whom seem to find inherently dubious the idea that the humanities disciplines *are* disciplines. Members of the general public (as represented, for example, by the city councils and state legislatures that fund most public higher education in the United States) commonly associate 'the humanities' with the reading and teaching of edifying books and are inclined, accordingly, to see humanities faculty as people who get paid for doing what anyone could do if they were not busy earning a living doing real work. Similarly, biologist Gross and mathematician Levitt confidently contrast the time, devotion and 'appallingly hard work' required to attain competence in the sciences with the unspecified but pointedly demeaned 'style of education and training that nurtures the average humanist'.[20]

The attitudes and assumptions in play here certainly reflect widespread public ignorance of the nature – or, indeed, existence – of specialised training and scholarship in the humanities. They also reflect a growing tendency within the academy itself to identify intellectual activity and achievement with the production of palpable, visible, measurable and more or less immediately applicable knowledge joined to a longstanding identification, in the academy as elsewhere, of the production of such knowledge with the natural sciences. There are, of course, more spacious ways to conceive intellectual activity and subtler ways to calculate its value. Moreover, the conception of the sciences just mentioned can be seen as dubious in every way – historically, conceptually and, indeed, practically. Such narrowly scientistic views of knowledge, however, along with quasi-positivist or vulgar utilitarian views of science are in better accord with the increasingly 'production'-centred ethos of many universities than the alternative views of both (that is, of knowledge and science) elaborated in constructivist epistemology or contemporary science studies. There is good reason to worry, then, that when limited resources are allotted among different sectors of the academy, Sokal's caper and other science-war battle shows will operate tacitly or be invoked explicitly as evidence for the intellectual decline and relative expendability of the humanities *and* for the absurdity of just such alternative views.[21]

A final important point here. The kinds of asymmetry just described in public understandings of the humanities vis-à-vis the natural sciences are commonly joined and amplified by the more general public partiality toward *traditional* views. That is why terms such as 'fashionable' or 'trendy' do so much work so cheaply. Like political incumbents, intellectual incumbents have a strong advantage over newcomers, and for many of the same reasons – greater name familiarity, more visible marks of authority, readier access to public platforms, more control over the procedures of institutional certification (degrees, titles, awards) and so on. Thus it is not surprising that rhetorically resounding, if intellectually thin, reaffirmations of traditional ideas (about science or anything else) find readier acceptance by the general public than challenges to those ideas, however worthy, or that few members of even the general *academic* public stay around long enough to hear the details of genuinely original views or attempt to explore them intellectually at any level. If we have no particular personal or professional stake in some question, we are generally happy to be told by some apparently relevant authority – a colleague in that field, a reviewer in our favorite periodical – that some radically new way of thinking about it is not worth our trouble. And if we do have a personal or professional stake in the question and the new way of thinking is manifestly contrary to our own way, then we may be all the more eager to hear about the low motives and questionable credentials of those who propose and support it. Hence, one suspects, the grim but perhaps also *reassuring* conviction among many academics that certain quarters of the university are being invaded by a gaudy troop of 'deconstructionists' and 'postmodernists' who have fled from reason, despise truth, promote self-evidently absurd doctrines and are attempting to destroy science. It is easier to accept *that*, it seems, than to undertake a first-hand study of the ideas in question or to give a serious hearing to the alternatives they offer.

V

In view of the range of intellectual and institutional problems indicated above, certain remedies suggest themselves. One would be broad attention to the design and teaching of two-culture-crossing courses that took seriously the interests of humanities students in the natural sciences and respected those students' general intellectual abilities and specific skills and knowledge. Another would be the creation of interdisciplinary research centres where natural scientists, social scientists and humanists had multiple, diverse occasions to interact intellectually over extended periods of time. Such projects, I think, are all to the good, and, where pursued, show, in my experience, encouraging degrees of success. I would like to conclude, however, with a few words on an ostensibly remedial project that has a strong following in some quarters of the academy but that is, I think, distinctly *not* to the good. I refer to the idea, currently made popular by E. O. Wilson's book, *Consilience*, that we should solve the problems of disciplinary difference by 'closing the gap' between the natural sciences and other fields of study so that all knowledge would be in fact what it should be and is clearly destined to become, namely, *unified*.[22] This is, I think, all pretty dubious.

First, what we speak of currently as the 'natural sciences', 'social sciences' and 'humanities' are, of course, only relatively stable clusters of continuously emerging, developing, combining and differentiating intellectual traditions and practices. Integrations and *dis*integrations of these kinds appear to be fundamental processes and phases of intellectual history.[23] Accordingly, the unification of all knowledge does not appear to be a plausible or even meaningful eventuality. It also does not appear to be a desirable ideal. Everything we know about the dynamics of intellectual history indicates that the play of differing – and indeed conflicting – perspectives is a necessary condition for the emergence of new ideas and practices in any field. In view of the tendency of all established conceptual systems to move toward self-affirming structures of ideas and of all disciplines – including the natural sciences – to be at risk of stagnation from taken-for-granted assumptions and habitual practices, the maintenance of this condition of epistemic multiplicity and divergence appears crucial for the continued vitality of any intellectual community. That is why one would promote the vigorous, ongoing interaction of different disciplines and disciplinary practitioners: the mutual appropriation of skills and techniques, the inter-translation of concepts and findings, the extension of models and theories into new domains of application, the provision of new perspectives on old problems and so forth. *Interaction*, however, is not *integration*, and certainly not *incorporation*. Moreover, *mutual* appropriation is a two-way street, involving an acknowledgement of the value of intellectual traffic – in ideas, perspectives, findings and theories – from the humanities and social sciences to the natural sciences as well as the other way around.

In Wilson's book and related proposals for the unification of knowledge, the sole measure of epistemic value is taken to be scientificity, understood along narrow, crudely logical-positivist lines. Accordingly, the interpretive social sciences, along with the humanities, are seen as simply *un*scientific or at best *pre*-scientific, and integrating the two cultures becomes largely a matter of absorbing the anarchic humanities and floundering social sciences into the more orderly and grown-up natural sciences.[24] The aims and attitudes reflected in such proposals are illustrated at some length in an essay, 'The Psychological Foundations of Culture', by evolutionary psychologists John Tooby and Leda Cosmides, who urge a 'conceptual integration' of the social and natural sciences. In a telling passage, the authors allude to the work of cultural anthropologists Edmund Leach and Clifford Geertz, whom they charge with 'abandoning the scientific enterprise' altogether, Leach by 'reject[ing] scientific explanation as the focus of anthropology' and Geertz by advocating 'treating social phenomena as "texts" to be interpreted just as one might interpret literature'.[25] Tooby and Cosmides continue:

> These positions have a growing following, but less, one suspects, because they have provided new illumination than because they offer new tools to extricate

> scholars from the unwelcome encroachment of more scientific approaches. They also free scholars from all the arduous tasks inherent in the attempt to produce scientifically valid knowledge . . . Those who jettison the epistemological standards of science are no longer in a position to use their intellectual product to make any claims about what is true of the world or to dispute the others' claims about what is true.[26]

The degree of complacent condescension here is extraordinary by any measure, but the charges or insinuations (turf-protection, intellectual laziness, retrograde resistance and so forth) are typical. To literary scholars, cultural anthropologists and others in the humanities or social sciences acquainted with the practice of 'treat[ing]' things 'as texts', the phrase suggests taking the phenomena in question (for example, cultural practices or creative works) as the products of expressive human subjects; examining and describing their forms, patterns, themes and internal relations; considering their relation to other phenomena of like kind or genre; attempting to elucidate their culturally conventional and/or otherwise expressive meanings;[27] and reflecting on their broader connections and implications – for example, historical, ethical or experiential. So understood, the phrase also suggests a disciplined activity involving specific aptitudes, skills and knowledge as well as intellectual labours of various kinds, some 'arduous' enough. For many scientists, however, and other academics trained in or strongly identified with a positivist tradition (for example, analytic philosophers or historical scholars), it evidently suggests idle, self-indulgent opinion-mongering.

From the perspective of Tooby and Cosmides, cultural anthropology in the mode of Leach or Geertz, in treating cultural objects and practices 'as texts' rather than seeking to identify their 'underlying causal mechanisms',[28] renders itself intellectually and critically worthless: its practitioners cannot claim to be revealing the truth nor can they 'use their intellectual product' to dispute the truth-claims of anyone else.[29] From other perspectives, however, it seems clear that identification of the underlying causal mechanisms of phenomena (where such mechanisms do, in fact, exist) is neither the only kind of knowledge there is nor the only kind worth having and, moreover, that the intellectual value of interpretive disciplines such as cultural anthropology, art history or literary criticism arises precisely from their 'treat[ing]' various things (including texts) 'as texts' and from their ability, thereby, to examine important aspects of human behaviour and experience. Divergent understandings and assessments of this kind are products, of course, of the very disciplinary segregations at issue here. One may, accordingly, applaud efforts to escape the intellectually confining consequences of such segregations, including, as discussed throughout this chapter, the mutual stereotypes and antagonisms they foster. These are not, however, either the evident motives or likely results of the integrationist projects under discussion here. On the contrary, those projects and the manner of their promotion are themselves the products of a conspicuously intransigent 'two cultures' ideology, glibly and grimly enacting its familiar complacencies, hostilities and provincialisms.

Returning one last time to our etic/emic chart, we could say that the 'gap' between 'the two cultures' is not a pathological condition to be healed but, rather, a natural enough outcome of the dynamics of human cognition or, indeed, of human ecology – that is, of the multiple ways in which we interact with our environments and the multiple epistemic stances to which they give rise. Contrary to the arguments of Wilson or Tooby and Cosmides, there is no reason to think there is some single all-purpose 'best' way for human beings to get to know their worlds or some ultimate state of full and totally integrated knowledge toward which all minor, imperfect and disparate knowledges are tending or should be marshalled. There is also no reason to think the human community would gain anything of intellectual or ecological value – truth, rationality, wisdom, individual effectiveness and productivity, group survival, or the preservation of the biosphere (to name some goals commonly cited) – if we banished all 'soft' (emic) understandings and open-ended ponderings and replaced them with 'hard' (etic) causal explanations and application-driven analyses, or, for that matter, if we declared a moratorium on technoscience. On the contrary, there is good reason to think that all these forms of knowledge-seeking, with all their internal differences and mutual frictions, are required for the maintenance and flourishing of our *natural* – which is also to say *cultural*, which is also to say *ethical, aesthetic* and *reflective*—relation to our environments, including each other and that which we have created.

Notes

1. Snow, *The Two Cultures*. An earlier version of this chapter was delivered at a symposium, 'The Two Cultures Revisited', held at the University of Dayton in February 2001.
2. For analysis of some of these developments, see Abbott, *Chaos of Disciplines*; Brint (ed.), *The Future of the City of Intellect: The Changing American University*; Newfield, *Ivy and Industry: Business and the Making of the American University, 1880–1980*.
3. Usage of the term 'humanist' has been and remains extremely various. It can refer to people who study and revere ancient texts; people who believe there are ethically potent bonds of commonality among all humans; people

who are partisan to human beings in opposition to deities, machines, animals, particular nations or particular ideologies; people who see a sharp division between natural phenomena and the acts or productions of human beings; and, most recently, people who train and teach in what are called 'humanities' disciplines in colleges and universities. It is only in this last sense that the term is used in this chapter.

4. B. H. Smith, *Belief and Resistance*.
5. Snow was a trained scientist and, though a writer of widely read novels, not a member of the British literary establishment. His conviction of the superiority of the 'culture' of science was explicit and his bias in its favour evident throughout his essay.
6. These definitions do not specify to whom the descriptions or interpretations are being given, which would affect how they were framed. For the purposes at hand, I assume that the anthropologist's 'etic' descriptions and explanations are addressed to his or her colleagues and not, at least not in the first instance, to a native, and that the native's 'emic' descriptions and interpretations are addressed to another member of the tribe and not, at least not in the first instance, to a visiting anthropologist.
7. An influential version of these ideas was put forth by Wilhelm Dilthey (1833-1911). See his *The Formation of the Historical World in the Human Sciences*, eds Rudolf A. Makkreel and Frithjof Rodi (Princeton: Princeton University Press, 2002 [orig. *Der Aufbau der geschichtlichen Welt in den Geisteswissenschaften*, 1910]). In alternative schemes and nomenclatures (as in this translation of Dilthey's work), 'the natural sciences' are distinguished from 'the human sciences'.
8. For examples and analysis, see Connolly, *Identity/Difference*.
9. See B. H. Smith, *Contingencies of Value*, pp. 36–42; *Belief and Resistance*, p. xvi.
10. On the distinct cultures of high-energy physics and microbiology, see Karin Knorr-Cetina, *Epistemic Cultures: How the Sciences Make Knowledge*. Fleck's account is discussed in Chapter 3 above.
11. For a relevant sociological account and analysis, see Bourdieu, *Distinction: A Social Critique of the Judgment of Taste*.
12. Although all these are egregious and pervasive in Western culture, they are not, of course, restricted to it.
13. See Collini, 'Introduction', p. xxvi.
14. Major figures involved were biologist Paul Gross, mathematician David Levitt, cultural critic Andrew Ross, physicist Alan Sokal and literary scholar/theorist Stanley Fish. For a rehearsal of some of the exchanges, see *The Sokal Hoax: The Sham that Shook the Academy*. For an extensive analysis and perceptive discussion, see Huen, *In the Wake of the Science Wars: An Experiment with the Anthropology of the Academy*.
15. Reactions of these kinds are provoked especially by Latour and Woolgar, *Laboratory Life* and Collins and Pinch, *The Golem: What Everyone should Know about Science*. For examples and discussion, see Labinger and Collins, *The One Culture?: A Conversation about Science*; Huen, *In the Wake*.
16. The provenance of such general terms as *force* or *evolution* and of many specialised terms such as *complementarity* or *incommensurability* is not

readily determinable and, of course, origination of a term carries no proprietary rights regarding its future usage.

17. See, for example, Gross and Levitt, *Higher Superstition*; Gross, Levitt and Lewis (eds), *The Flight from Science and Reason*; Sokal and Bricmont, *Fashionable Nonsense*; Koertge (ed.), *A House Built on Sand*. Gross and Levitt give little evidence of first-hand contact with the work of the philosophers and theorists (Foucault, Derrida and Latour among them) whose ideas they pronounce absurd and describe as 'cryptic rituals', 'convoluted cabalistic fantasies' and 'conceptual freak show[s]' (*Higher Superstition*, pp. 37, 8, 88). The sources for these pronouncements and descriptions, as indicated by the citations in *Higher Superstition*, consist largely of articles by uninformed, hostile journalists and academics with sizeable axes of their own to grind.
18. Academic philosophy is often the higher authority invoked in these battles in defence of traditional views. Thus the resistance of orthodox philosophers to the ideas of various historians and sociologists of science – or the fact that, as Gross and Levitt put it, their views 'drive earnest and responsible philosophers of science into paroxysms of disgust' (*Higher Superstition*, p. 139) – is offered as an argument against those ideas. But, of course, academic philosophy is not itself outside or above these debates since many of its aims, claims and procedures are directly challenged by the intellectual developments at issue.
19. Hofstadter, 'Popular Culture and the Threat to Rational Inquiry'.
20. Gross and Levitt, *Higher Superstition*, pp. 240, 5.
21. For cautionary examples and general discussion, see Karp, *Shakespeare, Einstein, and the Bottom Line: The Marketing of Higher Education*.
22. Wilson, *Consilience*.
23. See, for example, Oleson and Voss (eds), *The Organization of Knowledge in Modern America, 1860–1920*; Easton and Schelling (eds), *Divided Knowledge: Across Disciplines, Across Cultures*; Collins, *The Sociology of Philosophies: A Theory of Global Intellectual Change*; Abbott, *Chaos of Disciplines*. For sceptical views of the supposed unity or projected unification of the natural sciences, see Galison and Stump (eds), *The Disunity of Science: Boundaries, Contexts, and Power*; Bechtel, 'Unity of Science'. Bechtel writes:

    > Although the dreams of unity of science via reduction advanced by the positivists have generally not panned out, unification and integration, viewed in a more patchwork manner, are a routine part of modern science . . . The resulting picture is a network of local integration, not one of global unification. (p. 857)

24. Not all such proposals for remedial integration are from scientists. Thus various humanities scholars urge the wholesale importation of one or another high profile scientific (or, sometimes, scientoid) programme (for example, currently, cognitive science or evolutionary psychology) into their field with the stated aim of bringing intellectual respectability to its otherwise amateurish, impressionistic efforts (see, for example, A. Richardson and Steen (eds), *Literature and the Cognitive Revolution*; Carroll, *Literary*

Darwinism: Evolution, Human Nature, and Literature). Others in the humanities state their conviction that, to cure its presumptive ills or misfortunes, their particular field (literary study, art history, musicology and so forth) must *become* a science, a conviction commonly attended by a dubious – certainly limited – idea of what it is to be a science and by a visibly dyspeptic attitude toward new and currently favoured approaches in the field in question (see, for example, Guillory, 'The Sokal Affair and the History of Criticism').

25. Tooby and Cosmides, 'The Psychological Foundations of Culture', p. 22.
26. Ibid.
27. Diverse conceptions of *meaning* and accordingly of *texts* and *interpretation* are, of course, part of the issue here. Like a number of other evolutionary psychologists, Tooby and Cosmides owe their views of such matters largely to Noam Chomsky's referentialist, innatist, universalist linguistics (see, for example, Pinker and Bloom, 'Natural Language and Natural Selection'; Pinker, *The Language Instinct*) and would doubtless reject the structuralist, poststructuralist, and/or cultural- and linguistic-relativist views that inform the work of many cultural anthropologists.
28. According to Tooby and Cosmides, the cultural and social phenomena now improperly studied in the social sciences and merely interpreted in the humanities are the superficially variable expressions of underlying innate, universal mental mechanisms that evolutionary psychologists 'identify'. The considerable problems with these and other claims of evolutionary psychology are examined in Chapter 6.
29. Leach's disclaimer in that regard ('Social anthropology,' he writes, 'is not, and should not aim to be, a "science" in the natural science sense . . . Social anthropologists should not see themselves as seekers after objective truth' (Leach, *Social Anthropology*, p. 52)) is represented by Tooby and Cosmides as a scandalous confession of non-seriousness ('Psychological Foundations', p. 22).

Chapter 6

Super Natural Science: The Claims of Evolutionary Psychology

Evolutionary psychology, a recently constituted but already broadly extended programme in the study of human behaviour, is notable for, among other things, the unusually pre-emptive character of its claims. According to its major proponents, 'reverse engineering', the method that defines and distinguishes evolutionary psychology, permits identification of the underlying, innate mental mechanisms that govern all human behaviour, from incest-avoidance and female-adolescent anorexia to past-tense formation and a taste for Victorian novels. In supplying these identifications, it is said, evolutionary psychologists provide genuinely scientific explanations for human behaviours and cultural practices that, up to now, have been improperly or inadequately explained by other social scientists and, at best, merely 'interpreted' in the humanities. It is claimed, moreover, that, in thus furnishing the missing link between the natural and the human sciences, evolutionary psychology has effected a crucial turning point in intellectual history, inaugurating a conceptual integration of all fields of genuine knowledge under the mantle of a single, comprehensive scientific discipline or, in effect, a *super* natural science.[1] As I indicate below, there is good reason to be sceptical of these somewhat grandiose claims.

I

A bit of preliminary map-sketching will be useful. Contrary to depictions by evolutionary psychologists, the most significant controversies over the claims and accomplishments of the programme are not between enlightened Darwinists and dogmatic theologians or between sober, up-to-date cognitive scientists and either ideology-driven humanists or social scientists clinging to archaic ideas.[2] They are, rather, between practitioners and promoters of evolutionary psychology per se and scientists and theorists in related fields – for example, developmental psychology and paleoanthropology – who question the empirical and conceptual validity of the assumptions and methods of the new programme and maintain the value of other approaches to the study of human behaviour and cognition.[3]

In a number of respects, these controversies rehearse familiar conflicts between central but divergent traditions in philosophy of mind and philosophical psychology. These include, most explicitly, rationalism versus empiricism, mentalism versus physicalism, and nativism or innatism versus environmentalism. Each of these traditions and their mutual relations have undergone extensive modification and complication over the past 300 years, especially since the separation of empirical/physiological psychology from speculative philosophy at the end of the nineteenth century.[4] Now, at the beginning of the twenty-first century, the major alternatives to rationalist/mentalist/innatist accounts of human behaviour are not 'blank slate' empiricism or cultural determinism but various accounts and models ('dynamic' or 'ecological', as they are sometimes called) of the emergence of individual behaviours and species-typical traits and tendencies from the contingent interactions of humans and their environments – accounts and models, it should be stressed, that are extensively informed by and responsive to current research in, among other fields, evolutionary biology and cognitive neuroscience.[5]

Today, the rationalist/mentalist/innatist tradition survives mainly in analytic-rationalist philosophy of mind, computational-modularist accounts of cognition and current versions of Chomskian language theory. Evolutionary psychology can be seen as a selective grafting of elements of each of these three projects onto E. O. Wilson's original programme of sociobiology.[6] The resulting hybrid carries key features of its various forebears but also significant new features. Thus, whereas first-generation sociobiology explained the phenomena of human behaviour and culture as the expression of uniform, universal, genetically encoded *tendencies and capacities*, evolutionary psychology explains them as the expression of uniform, universal, genetically encoded *mental structures and processes*. According to Wilson et al. in the 1970s and 1980s, what makes us act the way we do is primarily our *genes*. According to Leda Cosmides, John Tooby, Steven Pinker et al. since the early 1990s, what makes us act the way we do is primarily our *minds*, but what makes our *minds* act the way they do is primarily our *genes*. The differences are both profound and negligible.

II

In examining evolutionary psychology here, I focus on its assumptions, methods and claims primarily as expounded by Barkow, Cosmides and Tooby in the inaugural collection, *The Adapted Mind*, and as popularised by Pinker in his 1997 book, *How the Mind Works*.

To pose the question of human cognition in the terms of the title of Pinker's book is, of course, already to presuppose a good bit of the answer: first, that the mind is a discrete and readily distinguishable entity; second, that it operates mechanically; and third, that its operations could be the object of a causal, naturalistic explanation. Each of these suppositions can be and has been questioned. Thus 'the mind' may be seen as a name given to a historically shifting set of heterogeneous phenomena and notions, ranging from introspected experiences and observable patterns of behaviour to the various faculties, processes and interior mechanisms that, at various times and in various informal and formal discourses (philosophical, ethical, legal, medical and so forth), have been posited or assumed to explain them. Accordingly, the mind as such may not appear to be the sort of thing that cognitive scientists – as distinct from, say, intellectual historians – should seek to explain or ever could quite 'explain' at all.

Pinker is not unaware of such views, but dismisses them early on,[7] and handles the questions they raise through facile definitions or ad hoc slides between 'mind' and 'brain' or between each of these and 'mental organs', 'neural circuitry', 'our thoughts and feelings' and much else besides. The conceptual problems thereby evaded, however, are significantly implicated in each of the two approaches that Pinker and evolutionary psychology more generally seek to synthesise – namely, a strictly computational model of mind and a narrowly adaptationist account of human behaviour – and return, with other problems, to unsettle the ambitious claims made for the new discipline thus constituted.

Strategically equivocal on the relation between 'the mind' and 'the brain', Pinker neither quite identifies nor quite distinguishes between them. Thus, after declaring that 'the mind is . . . what the brain does',[8] he explains 'mental organs' as both complements of and analogues to 'bodily organs'. But if the mind is what the brain *does*, then it is not clear why we need 'mental organs' at all. It would seem to be the same as saying that birds have 'flying organs' as complements of and analogues to their wings.[9] At other points, Pinker identifies mental organs with hardwired (but as yet undetected) neural circuits, which would seem to make them more a matter of how the brain is constructed than of what it *does*. My concern here is not only the equivocal usage as such but its signalling of a more fundamental difficulty in the project of evolutionary psychology, namely, its effort – Quixotic, in my view – to provide a quasi-naturalised account of traditional rationalist conceptions of cognitive processes and, accordingly, its continuous need to finesse the mind/body problem.

Comparable switches and slides recur in *The Adapted Mind.* For example, in their introductory essay to the volume, Barkow et al. write: 'The brain takes sensorily derived information from the environment as input, performs complex transformations on that information, and produces either data structures (representations) or behaviour as output.'[10] Although reference is made here to what the *brain* does, the authors go on to stress the non-neurophysiological and apparently non-physical nature of the mechanisms involved: '[A neuroscience description] tells only how the physical components of the brain interact'; 'a cognitive . . . description [such as that offered by evolutionary psychology] characterizes the "programs" that govern its operation.'[11] The contrast is characteristic,[12] but cuts two ways. First, the distinction between physical embodiment and mental processes permits evolutionary psychology to claim autonomous status as a discipline, free to deduce mental organs and programmes with minimal constraints from such ('only . . . physical') considerations as neurophysiology. Conversely, however, in marking off the *mental* as the specific territory of their accounts, the authors reveal the paradoxical character of the new field, which claims a scientificity superior to that of other social sciences but produces, as its central explanatory resource, a distinctly non-empirical realm of causality drawn from a distinctly non-empirical tradition of philosophical psychology. Significantly, in this seemingly even-handed division of disciplinary labours ('physical components of the brain' or, as Pinker and others often put it, 'hardware' to neuroscience, 'governing "programmes"' or 'software' to evolutionary psychology), the authors ignore a major project and achievement of contemporary neuroscience – namely, the correlation of observable neural structures and events with both observable behaviours and reported subjective experiences, and, crucially, the explanation of such correlations in terms of specific neurophysiological and other bodily processes without the postulation of intervening mental mechanisms.[13] Indeed, in view of such ongoing work in cognitive neuroscience, it appears that, whatever else may be said about the mind, insofar as it is conceived as a realm of *interior mediating causality*, it remains a totally hypothetical and arguably superfluous construct.

Describing the methodological centrepiece of the new discipline, 'reverse engineering',[14] Pinker observes that, just as in the case of an artifactual contrivance (his example is an olive pitter), so also in the case of the mind: we can explain how it works only by – and in – identifying the purpose for which it was 'designed' and then, on the basis of other assumptions, deducing how it must have been engineered to achieve that purpose.[15] Though psychologists, Pinker informs his readers, have been unable even to approach this task properly in the past (the history of the discipline being represented, accordingly, as a series of bumblings and bungles), current practitioners of evolutionary psychology, by virtue of their rigorous adherence to two key ideas, have now almost completed it. The first is the idea that the mind, like a computer, is an information-processing machine, the purpose of which is to solve problems through rule-governed manipulations of symbols that represent objective features of the world. The second is the idea that the mind, like the body, consists of multiple individual organs or 'modules' engineered by natural selection to maximise the reproductive fitness of our Upper Paleolithic ancestors and reflecting that design more or less directly in their current operations.

In his account and defence of the method, Pinker represents as a sequence of logically linked productive inferences what could otherwise be seen, both from his own examples and from the descriptions and explanations of Tooby and Cosmides, as a process of self-enclosed speculation directed by a set of mutually determining, mutually validating assumptions, descriptions and hypotheses.[16] For example, having first described vision as the brain's computational solution to 'the problem' of the mind's need to obtain 'accurate knowledge of the world' from the 'shifting, impoverished' patterns of light registered by the eyes, Pinker goes on to explain how evolutionary psychologists deduce the existence of neural circuitry designed by natural selection to perform computations to solve just that problem.[17] The putative 'problem' here, however, is an artifact of the very computational model of mind that evolutionary psychology takes as its initial premise. For it is only in accord with the realist/representationalist assumptions of that model that visual perception would be seen as 'defective' informational input in the first place, or that the brain would be seen as requiring 'correct representations' of the world in order to direct behaviour appropriately. Other current models of cognition that describe visual (and other) perception without such ontological and epistemological presuppositions offer accounts of appropriate (that is, self-sustaining and effective) organism-environment interactions without the postulation or 'deduction' of extensive preloaded computational software.[18] Evolutionary psychology appears to be a theory that yields a method of analysis that generates the solution of problems created by the theory itself. Of course, it has some good company in those respects.

III

Proponents of evolutionary psychology often secure a spurious sense of authority for key assumptions of the programme by effacing important alternative ideas or by conflating them with readily discountable adversaries, such as obsolete vitalism, politically correct 'social constructionism', or preposterous 'postmodernism'.[19] Contrary to the impression thus created, however, the central assumptions of evolutionary psychology remain controversial within the relevant natural science fields and among cognitive scientists more generally.

Among these assumptions is the computational model of mind – that is, the idea that the mind, like a computer, is an information-processing machine designed to solve particular, discrete, externally-posed problems in accord with hardwired rules. Criticisms of this model of mind, itself developed largely by engineers, logicians and mathematicians, often reflect specifically *biological* considerations. Thus, biological-systems theorists and developmental psychologists note that prevailing versions of the model (other more or less radically modified versions, including 'connectionist' and 'neural network' models, continue to emerge)[20] do not adequately reflect the peculiar structural and operational properties of living systems, including their characteristically global, self-regulating and, in important respects, nonlinear dynamics.[21] It has also been observed that, in embodied, mobile, socially embedded and verbal creatures such as ourselves, cognitive processes involve complex forms of social, perceptual and manipulative coordination as well as internal feedback mechanisms that, again, have no counterparts in the computational model as currently developed.[22] While Pinker attempts to discredit such criticisms by associating them with vitalism (life as 'a quivering, glowing, wondrous gel'),[23] it is not a matter of honouring some ineffable distinction between organisms and physical systems but of understanding what kinds of physical systems organisms – including human beings – are.

Major alternative theoretical frameworks for explaining intelligent human behaviour and cognitive processes more generally give due weight to the relevant characteristics of living systems. Thus the processes of cognition in humans (as in organisms more generally) are seen not as a series of discrete problem-solving computations but, rather, as the continuous modification of our structures and ways of functioning by, and in the course of, our ongoing interactions with an always changing environment.[24] Moreover, as observed by theorists of cognition for the past half-century, human-like intelligence includes embodied skills that cannot be reduced to or translated into statement-like propositions, manipulations of symbols or the types of algorithmic reasoning commonly favoured in the field of artificial intelligence.[25] Similarly, perception is widely understood not as the passive reception of environmental 'inputs' but as an ongoing activity involving the entire organism in its effective negotiation of various features of its environment.[26] Computational models of mind can explain human behaviour or cognition only to the extent that ongoing activities and processes can be decomposed into sets of context-free manipulations of discrete bits of information in accord with prior and fixed rules. Such explanations may be plausible for activities such as playing chess or calculating taxes – that is, for the sorts of human-like problem-solving activities that computers have been programmed to accomplish more or less successfully. But activities of that sort constitute only a fraction of what might reasonably be understood as intelligent behaviour, whether in humans or other organisms or, for that matter, in (other) machines.

Other criticisms of the computational model involve broader conceptual, including epistemological, considerations. For example, our individual relations to our environments can be seen as a matter of mutual determination and transformation rather than, as in the computational model, unidirectional informational inputs and problem-solving outputs. In the case of humans as, again, all other organisms, a creature's environmental niche or operative world – what it can interact with perceptually and behaviourally as distinct from what an observer may notice as impinging on it – is specified by that creature's particular structure and modes of operating, and these latter are modified in turn by the creature's ongoing interactions.[27] A human being, a cat and a paramecium, even if they share the same physical space, do not live out their lives in the same environment. Accordingly, the realist assumptions that the computational model takes over from classic philosophy of mind, in which '*the* [single, given] environment' is conceived in terms of a set of autonomously determinate features, can be seen as crucially confining or, indeed, disabling.

Dubious epistemological and related ontological presumptions are especially evident in the accounts of category-formation offered by evolutionary psychology. Thus Pinker assures readers that categories are not 'arbitrary conventions that we learn along with the other cultural accidents standardized in our language' (a view, certainly curious as stated, that he attributes to 'many anthropologists and philosophers'), but are 'forced' on us, via innate mental feature-detectors, by the way the world actually is.[28] 'Mental boxes work,' he writes, 'because things come in clusters that fit the boxes.'[29] Or, Pinker observes at another point (the issue is social categorisation, as exemplified by the homosexual/heterosexual binary), 'the dichotomy between "in nature" and "socially constructed" . . . omits a third alternative: that some categories are products of a complex mind designed to mesh with what is in nature'[30] – which is, of course, hardly a third alternative (though such do exist) or an escape from that dichotomy.[31] Pinker's allusion here to a 'mesh' between mind and world signals another central assumption of evolutionary psychology, this one carried over from the classic idea of the mind as a 'mirror of nature'. Thus Tooby and Cosmides observe, similarly, that 'there has been the evolution of a mesh between the principles of the mind and the regularities of the world, such that our minds reflect many properties of the world'.[32] No evidence is offered for the assertion, and it is no easier now than it has been for the past two or three centuries to imagine what evidence *could* be offered for it.[33]

Pinker's related account of concept-formation is exceedingly strained and comparably dubious. '[A] handful of concepts about places, paths, motions, agency, and causation,' he writes, 'underlie the literal or figurative meanings of tens of thousands of words and constructions' in all languages.[34] The widespread recurrence of certain elementary ideas and relational concepts is generally granted; the question, however, is how they come into being.[35] According to Pinker, they are innate and prewired, making up a universal conceptual lexicon – 'mentalese' – the elements of which combine with input from a local physical environment to form complex thoughts according to combinatorial rules that are also prewired.[36] Connections between our ancestors' eyes, muscles and neural circuits, Pinker tells readers, were originally formed by natural selection 'for reasoning about rocks, sticks' and about 'space and force' but were subsequently 'severed', and 'references to the physical world were bleached out', so that the 'slots' (neural? mental?) thus left empty could be 'filled in with symbols for more abstract concerns like states, possessions, ideas, and desires'.[37] An ingenious story. There is, however, no evidence either for the existence of any such concept-specifying prewiring or for the occurrence of any such severing or bleaching-out processes – at least not in any of the relevant empirical fields, such as neurophysiology or paleoanthropology. Indeed, like many other reverse-engineering deductions, the only ground for supposing the existence of such structures and processes is their explanatory necessity, given a prior commitment to nativist/rationalist conceptions of human cognition and a determined neglect, dismissal or foreclosing of other avenues of research and theory.

Not surprisingly, analogies from the operations of computers figure largely in the computational model of mind. Reference to computers, Pinker maintains, 'demystifies' – but, significantly, also 'rehabilitates' – commonsense beliefs about the mind by supplying 'hard-nosed' materialist, physicalist explanations of 'fuzzy' mentalistic concepts.[38] To illustrate the point, he supplies a series of supposedly strict, precise translations: 'Beliefs are inscriptions in memory, desires are goal inscriptions, thinking is computation, perceptions are inscriptions triggered by sensors, trying is executing operations triggered by a goal.'[39] How much illumination is thereby provided, however, is questionable. 'Goal inscriptions' and 'inscriptions in memory' are, of course, no more material or physical (and no less reified) than 'desires' or 'beliefs', and none of these redescriptions is any more intrinsically 'hard-nosed' than the warehouse of other metaphors for cognitive processes drawn in the past from such technologies as agriculture, writing (the 'inscriptions' are obviously still there in force, though now recirculated through computers), hydraulics or cinematography. What Pinker exposes here, rather, is how a computational model of mind can give contemporary outfitting to traditional ideas and explanations without disturbing in the slightest their definitive – and, arguably, most problematic – features.

IV

Although evolutionary psychology claims scientific authorisation from evolutionary theory, its interpretations and appropriations of Darwinism are dubious along a number of lines. First of all, for the procedures of reverse engineering to make sense in relation to the principles of natural selection, every mental organ thereby deduced would have to be the end product of a series of genetic variants each of which had conferred an adaptive advantage on members of the species possessing it. While evolutionary sequences of that kind can be reconstructed for some clearly identified human capacities or traits (for example, our ability to recognise individual faces), in the case of the complex behavioural traits and often totally putative mental capacities posited by evolutionary psychologists (for example, a feature-detection device that guarantees the veridicality of the categories we form, or a mental module for syntactic parsing), it is at best difficult to construct, and often impossible even to hypothesise, any such scenario that would be simultaneously genetically, neurophysiologically and ecologically plausible. Thus, as a number of evolutionary theorists have pointed out, the explanatory accounts of evolutionary psychology, like those of its most immediate ancestor, sociobiology, require the assumption of prehistoric scenarios that range from the speculative to the unimaginable.[40]

Related objections can be raised to the unremittingly purposive, rational idiom of evolutionary psychology: 'genes for', 'designed for', natural selection as the ever-ingenious 'engineer' of fitness optimising 'devices' and so forth. What is obscured here is the significance in both classic and contemporary Darwinian theory of historical contingency – that is, not only the randomness of genetic mutations or the possibility of rare accidents, such as comets falling to earth, but the everyday chanciness, particularity and dependency of all biological events, developmental as well as evolutionary. The problem is not the technically imprecise language but the crucial explanatory oversimplification. If, as Pinker complains, people who stress the *interactive* dynamics of individual development have a 'phobia of ever specifying the innate part of the interaction',[41] it is not – or not always and only, as he implies – because of political anxieties but because, especially with respect to the complex or highly mediated traits usually at issue (for example, human intelligence, language use or sexual preferences and behaviour), 'specifying the innate part' is precisely what cannot be done.[42]

Pinker and other evolutionary psychologists have some serious phobias of their own, mostly focused on reminders of cultural and historical variability and the significance of individual experiences, social practices and normative institutions in shaping human behaviour. Invocations of such matters elicit Pinker's special scorn, and he can hardly find terms derisive enough ('myths', 'Romantic nonsense', 'anthropological correctness', 'the conventional wisdom of Marxists, academic feminists and café intellectuals' and so forth) to characterise them.[43] He is especially perturbed by the idea of *learning* (which he identifies, improperly and tendentiously, with mind-as-blank-slate empiricism)[44] and goes to great and often tortuous lengths to avoid acknowledging that our perceptual and behavioural tendencies – and, it may be presumed, neural circuits – are modified (strengthened, weakened, configured and reconfigured) throughout our lives by our interactions with our environments. For example, the discovery by developmental psychologists that stereoscopic vision is not present at birth, develops only gradually and is not always achieved does not challenge, for Pinker, the idea of its being a specific prewired capacity. Rather, it reveals for him the existence of a further supplementary mechanism, an alleged 'installation sequence' that programmes the (pre)wiring to occur *after* birth. While such a process, he acknowledges, 'requires, at critical junctures, the input of information that the genes cannot predict', it is, he believes, 'a better way of thinking about it' than 'the tiresome lesson that stereo vision, like everything else, is a mixture of nature and nurture'.[45] It is hard to see, however, in what way that lesson – however crudely stated here by Pinker[46] – is thereby controverted or bettered. Indeed, the significance of, in effect, *post-natal* neural assembly is a key point of the ecological and developmental-systems models mentioned above, models that Pinker conflates (under the label 'the "interactionist" position') with simplistic nature-plus-nurture views and dismisses exasperatedly (and without specific citation) as 'useless', 'a colossal mistake', 'ideas that are so bad they are not even wrong', and so forth.[47] Of course, he does not recognise his own duplication of that point here because, in the alternative interactionist models, the process(es) in question would be understood not as a fixed sequence (pre- plus post-natal) of discrete hardware-installations for receiving informational inputs from the environment but as the ongoing environmentally responsive reconfiguration of systemic connections throughout an organism's lifetime. There is, moreover, a further disturbing possibility that Pinker might well wish keep at bay. For if the emergence of such an obviously advantageous competence as stereoscopic vision can be explained in this way (that is, as partly a matter of contingent epigenetic development or, in effect, learning), then it is not clear why many other cognitive capacities should not be, from cheater-detection to past-tense formation and, indeed, the whole array of handy human abilities that evolutionary psychology insists are specifically innately prewired.

V

The intellectual confinements that result from Pinker's strenuously nativist presumptions are especially evident in his accounts of human social behaviour. While striking out at 'social construction[ism]' and 'fashionable "liberation" ideologies like those of Michel Foucault',[48] he evidently fails to grasp how some of the most elementary and ubiquitous instruments of human socialisation operate: for example, cultural representations such as images and lore, social sanctions such as shaming, and normative classifications sustained by institutions such as law or religion.[49] He certainly underestimates their naturalising force. Thus, in support of his conviction that a biologically programmed repugnance rather than a culturally transmitted taboo 'explain[s] what keeps siblings apart', Pinker observes without blinking: 'Brothers and sisters simply don't find each other appealing as sexual partners . . . the thought makes them acutely uncomfortable or fills them with disgust.'[50] *All* brothers and sisters? There is reason to think otherwise. *Simply*? Hardly.

The accounts of human social and sexual practices generated by evolutionary psychologists – from philanthropy and hypocrisy to courtship and spousal abuse – draw largely on concepts and related analyses developed by population geneticists and game theorists: 'inclusive fitness', 'reproductive investments', 'defections', 'payoffs' and so forth. In accord with such analyses, it is possible to calculate how certain behavioural choices (for example, cooperation rather than defection, submission versus flight or attack) would, under various hypothetical (and typically highly idealised) conditions, maximise the resources or enhance the reproductive fitness of the individuals pursuing them. Zoologists and behavioural ecologists use such calculations routinely to help explain a wide range of behaviours observed in other species. The idea that such analyses might also help explain various features of human social behaviour is not, in my view, mistaken in itself. What *is* mistaken is the idea, played out on page after page of *How the Mind Works* and other works of evolutionary psychology, that there is something especially rigorous and scientific about transferring such calculations and analyses as rawly as possible from blackboard, barnyard, jungle or presumed ancestral savanna to contemporary human society.[51]

'The human mating system,' Pinker writes,

> is not like any other animal's. But that does not mean it escapes the laws governing mating systems, which have been documented in hundreds of species . . . For human sexuality to be 'socially constructed' and independent of biology, as the popular academic view has it, not only must it have miraculously escaped these powerful [selection] pressures, but it must have withstood equally powerful pressures of a different kind.[52]

We may note in passing Pinker's distorted report of the idea of the social construction of sexuality and his related conflation of 'human sexuality' with 'gender roles' in the passage quoted below.[53] The more crucial and revealing point here is his evident presumption that the social behaviours at issue have *not* withstood any such pressures or could have done so only 'miraculously'. The passage continues: 'If a person played out a socially constructed role, other people could shape the role to prosper at his or her expense.' Well, yes, that *could* happen. Indeed, it is a pretty good summary of the relations between the more powerful/dominant and less powerful/dominant members of the species throughout recorded human history (one thinks readily of feudalism, slavery or literal patriarchy). Pinker's implication here, however, is that such shaping did not and could not happen. His argument and supposedly clinching counterexample go as follows: 'Powerful men could brainwash the others to enjoy being celibate or cuckolded, leaving the women for them. Any willingness to accept socially constructed gender roles would be selected out, and genes for resisting the roles would take over.'[54] Period. *QED*. Pinker evidently believes – and believes that he has just proved – that human beings have an innate, naturally selected resistance to the social construction of gender roles or, indeed, given the logic of his argument, to the social construction of *any* roles.[55] Since, however, the proof depends on an obliteration of the contrary evidence of recorded human history (as distinct from the compliant evidence of unrecorded Upper Paleolithic history) and of the actual range of behaviours of contemporary human beings (as distinct from what Pinker believes 'any bartender or grandmother' will tell you),[56] all it demonstrates are the explanatory confinements of evolutionary psychology and the perceptual biases of at least one of its major proponents.

VI

In an extended theoretical essay on the relation between the biological/genetic and the cultural, Tooby and Cosmides argue that the behaviours, patterns and practices that anthropologists and other theorists refer to as 'culture' and depict as socially transmitted and historically and otherwise variable are, in fact, merely the surface expressions of innate, uniform, genetically transmitted mental mechanisms designed to respond 'appropriately' to particular environmental inputs. They write: '[A]ny time the mind generates any behaviour at all, it does so by virtue of specific generative programmes in the head, in conjunction with the environmental inputs with which they are presented.'[57] Accordingly, cultural practices such as language, marriage customs or body decoration that appear to be socially transmitted among the members of a community are actually 'caused' by a mental process, 'inferential reconstruction', that Tooby and Cosmides describe as follows:

> This subset of cultural phenomena is restricted to (1) those representations or regulatory elements that exist originally in at least one mind that (2) come to exist in other minds because (3) observation and interaction between the source and the observer cause inferential mechanisms in the observer to recreate the representations and regulatory elements in his or her own psychological architecture.[58]

In support of this oddly prescriptive ('this subset . . . is restricted to') idea of cultural phenomena and evidently disembodied idea of cultural transmission, they argue: 'Other people are usually just going about their business as they are observed, and are not necessarily intentionally "transmitting" anything.'[59] Contrary to the implication here, however, nobody – certainly not anthropologists or most other cultural theorists – maintains that cultural transmission is necessarily or even primarily a matter of *intentional instruction*. The most significant alternative view, misrepresented or obscured here, is that people learn the practices of their groups the way they learn most other things: that is, through the modification of their behaviour in the course of their ongoing physical and perceptual interactions with their environments, including their physical, perceptual and other (for example, verbal) interactions with other people.

Tooby and Cosmides continue their description of this strictly mental sequence of events ('inferential reconstruction') with a set of terminological substitutions evidently designed to remove any suggestion of social communication and to stress the operation of innate mechanisms in individual receiving minds. 'Rather than calling this class of representations "transmitted" culture,' they write, 'we prefer terms such as *reconstructed culture*, *adopted culture*, or *epidemiological culture*.'[60] Such substitutions, however, simply beg the question and also reveal the strenuously confined and peculiarly interiorised character of their conception of the processes at issue. Though Tooby and Cosmides refer (in the passage quoted above) to *interaction* between 'source' and 'observer', the interactive phase of the sequence they describe consists entirely of the apparently automatic conveyance of discrete bits of information from one mind to another mind. No mention is made either of any *embodied* interactions, such as one person touching, physically leading or passing an object to another, or of the *mutually coordinated* interactions that, in the view of many cognitive and cultural theorists, are crucial to the social shaping and maintenance of complex human behaviours – interactions, for example, between parent and infant during early speech exchanges, or among children during play, or among members of a group during such collaborative activities as house-building or navigating. Nor is there any mention of the world of artifacts, tools, texts and other forms of material and symbolic culture which, along with acquired skills and routine practices, constitute the cultural scaffolding on which, it appears, much or perhaps all human cognition depends.

These omissions are striking and may seem perplexing, but can be understood in relation to the intellectual lineages of evolutionary psychology. Strongly indebted here as elsewhere to Noam Chomsky's account of language, the account of culture offered by Tooby and Cosmides adopts Chomsky's characteristic two-story surface/depth, performance/competence model along with other appearance/reality, exterior/interior dualisms familiar from classic rationalist thought. Thus at the upper, superficial level are all the observable phenomena, such as speech or marriage customs, represented as ephemeral and variable, while below, at the deep level, are the supposed basic and uniform but unobservable 'rules' or 'devices' that 'generate' or 'cause' them. Accordingly, and in conformity again with classic rationalist as well as Chomskian conceptions, an adequate explanation of the phenomena in question consists of 'discerning' or 'describing' – which is to say, positing – underlying structures and processes. The surface/depth model and related claims of discovery and description are evident in Tooby and Cosmides' account of why culture can be explained only through due study of its 'psychological foundations':

> To discern and rescue this underlying universal design out of the booming, buzzing confusion of observable human phenomena requires selecting appropriate analytic tools and frames of reference.[61]

> Only after the description of this [mental] architecture has been restored as the centerpiece of social theory can the secondary . . . effects of . . . social dynamics be . . . analyzed.[62]

The assumption that guides the selection of those analytic tools and frames of reference and determines them as 'appropriate' is that there *is* such an underlying universal design. Not surprisingly, what is thereby 'discerned', 'rescued' and 'restored' is the continuous reaffirmation of just that assumption.

Referring to the analytic method that results ('reverse engineering', as discussed above, or, in their terms, 'Evolutionary Functional Analysis'), Tooby and Cosmides observe: 'When they are linked together, the five components [of the analysis] not only provide a framework for the explanation of facts that are already known; they also form a powerful heuristic system for the generation of new knowledge.'[63] The method certainly generates something. The question is whether it is new knowledge or anything else intellectually substantive. Tooby and Cosmides defend the 'heuristic' operation of the method by way of an analogy to what is called 'functionalist analysis' in biology: that is, the effort to establish what biologically significant function (for example, food digestion or predator detection) a given species-typical feature (organ or behaviour) serves or may have served in the history of that species. They remark: 'Functionalist analysis in biology has motivated thousands of predictions about new and critical phenomena, whose subsequent discovery confirmed the productivity of the emerging paradigm.'[64] That, indeed, is the way it happens in biology. But it is not the way it happens or could happen in evolutionary psychology. Prediction and subsequent discovery of observable phenomena – organic features, the performance of biologically significant functions, paleological evidence for the past existence of either of these, and so forth – are not the same as prediction and alleged discovery of unobservable mental mechanisms. In effect, nothing is discovered in evolutionary psychology and no new factual knowledge is established. There are only increasingly elaborate structures of mutually presupposing and mutually supporting assumptions and hypotheses – the rhetorical-alchemical transmutation of sequences of linked *might*-be's and *would*-be's into confidently asserted *must*-be's.[65]

Evolutionary psychology advertises itself as a properly progressive natural-plus-human-sciences discipline or, as suggested above, a *super* natural science. Given the irremediably speculative and often exceedingly conjectural, not to say spectral, nature of its characteristic assumptions, methods, posits and explanations, one could view the new discipline as super-natural in other senses as well.

VII

The descriptions and historical accounts of the current intellectual scene that attend the self-promotions of evolutionary psychology are, as noted earlier, often tendentious. Thus Tooby and Cosmides maintain that the present 'isolation' of 'the human sciences' from 'the body of science' is the result of intellectually self-serving and regressive efforts on the part of certain 'scholarly communities' to retain a dualistic (spiritual/physical, human/non-human) view of the universe and its study.[66] While the natural sciences, they write, are 'becoming integrated into an increasingly seamless system of interconnected knowledge',[67] '[t]o many scholarly communities, conceptual unification became an enemy, and the relevance of other fields a menace to their freedom to interpret human reality in any way they chose'.[68] The historical romp and disagreeable insinuations are par for the course and, here as elsewhere, involve ignorance or obliteration of a substantial range of relevant institutional conditions, intellectual issues, and individual motives.[69] Nineteenth-century theorisations of distinctive types of knowledge did often draw on classical dualisms and theological notions of human exceptionalism and were accordingly limited. But they also expressed the conviction, hardly naïve or antiquated, that the methods and achievements of the *Geisteswissenschaften* – pursuits such as classical philology or literary exegesis – reflected aims and principles that, while different from those of the sciences or 'natural philosophy', were nevertheless intellectually substantive and significant. In the case of the newly emergent social sciences, notably anthropology and sociology, an insistence on disciplinary distinctions (for example, from individualistic psychology) reflected not, as Tooby and Cosmides darkly charge, a desire to escape the rigours of scientific standards or a refusal to grant the existence or relevance of biological or psychological mechanisms, but the idea – crucial for the respective identities of those fields and their rationales as discrete disciplinary programmes – that the regularities observable in social and cultural practices (for example, initiation rituals or the giving away of goods in potlatch) cannot be accounted for in terms of the beliefs, desires, or behaviours of *individual agents*.[70] In the case of the humanities, current self-distinctions from the natural sciences reflect, among other things, the continued power of the idea that *articulating* and *understanding* the world of *human experience* is irreducibly different from *describing* and *explaining* the phenomena of *human behaviour*. The idea of such an irreducible difference may be questioned in various ways.[71] It is improperly represented, however, as a wish on the part of non-natural-scientists to retain 'the freedom to interpret human reality in any way they chose'.

Contrary to the accounts given by Pinker, Wilson and Tooby and Cosmides and rehearsed by their followers,[72] the issues that divide evolutionary psychology from its major critics are not whether biology in general or evolutionary theory in particular is relevant to the study of human behaviour, or whether some and perhaps many endemic human traits or tendencies could be naturally selected adaptations. Both of these claims, where properly stated and duly understood, are commonly granted. The issues are, rather, whether the particular explanations offered by evolutionary psychology are intellectually cogent and can properly lay exclusive claim to due scientificity in the study of human behaviour and cognition; and, no less significantly, whether such explanations trump the understandings of the human scene, interior and exterior, developed by myriad other social scientists (ethnographers, psychologists, sociologists and so forth) over the past century and by myriad chroniclers (historians, biographers, diarists, diplomats, journalists, travellers, essayists, poets, novelists, playwrights and so forth) over the past two or three millennia. To dismiss all this as the theoretically benighted record of superficial phenomena or the worthless impressions of amateurs is not to create a rigorous science of human behaviour. It is to operate myopically in an impoverished universe.

Evolutionary psychology wants to call the mind-, culture- and cognition-explaining game over and to declare itself and its team the winners. Contrary, however, to its portrayal of the scene, the remaining questions in cognitive and behavioural science are not just technical, a matter of working out the details of a programme that all enlightened practitioners endorse. Quite the reverse: both fields are exceptionally active at all levels – conceptual, empirical and methodological – and also diverse and volatile, with new disciplinary configurations and domains of research opening up virtually continuously and significant ideas and connections being developed on all sides.[73] Like comparably ambitious projects from general equilibrium economics to structuralist linguistics, evolutionary psychology mistakes its own oversimplifications for the discovery of simplicity and its effacement of contingency, mediation and variability for the disclosure of universal laws. Due acknowledgement of the cultural, historical and individual variables that Pinker and Cosmides and Tooby dismiss would not compromise the development of naturalistic explanations of human behaviour and of whatever other phenomena 'mind' is taken to include. Similarly, due attention to the concepts, methods and findings of cultural anthropology, social history, sociology and related fields would not make evolutionary psychology less 'hard-nosed' than it is at present. On the contrary, such attention might introduce an order of observational precision, documentary concreteness and conceptual reach and subtlety (one thinks, for example, of Mary Douglas's classic analyses of pollution or of Erving Goffman's fine-grained studies of social interaction) sorely missing at present from the accounts of human practices produced in that field. Since the major current proponents of evolutionary psychology appear captive to an unregenerate two-cultures mentality, with its familiar intellectual provincialisms and disciplinary antagonisms, they are unlikely to see the value of such interdisciplinary possibilities.[74] As the vital and protean fields of cognitive and behavioural science continue to define and configure themselves, however, it may be hoped that new generations of scientists, scholars, and theorists will energise their development in such directions.

Notes

1. For these claims, see esp. Barkow et al., 'Introduction: Evolutionary Psychology and Conceptual Integration'; Tooby and Cosmides, 'The Psychological Foundations of Culture'; Pinker, *How the Mind Works* and *The Blank Slate: The Modern Denial of Human Nature*; E. O. Wilson, *Consilience*. For other representative expositions and applications, see Crawford and Krebs (eds), *Handbook of Evolutionary Psychology: Ideas, Issues and Applications*; Plotkin, *Evolution in Mind: An Introduction to Evolutionary Psychology*; Thornhill and Palmer, *A Natural History of Rape: Biological Bases of Sexual Coercion*; Buss, *The Dangerous Passion: Why Jealousy is as Necessary as Love and Sex*; Carroll, *Literary Darwinism*.
2. See Pinker, *Blank Slate*, pp. 5–13, and the descriptions of the so-called Standard Social Science Model in Tooby and Cosmides, 'Psychological Foundations', esp. pp. 24–34.
3. See, for example, paleoanthropologist Tim Ingold, 'Evolving Skills'; developmental psychologist Annette Karmiloff-Smith, 'Why Babies' Brains Are Not Swiss Army Knives'; philosopher of biology Paul E. Griffiths, 'From Adaptive Heuristic to Phylogenetic Perspective: Some Lessons from the Evolutionary Psychology of Emotion'; evolutionary theorist

Robert C. Richardson, 'Evolution without History: Critical Reflections on Evolutionary Psychology'. Critiques of specific claims and features of evolutionary psychology will be cited below where relevant.

4. For a good account, see Reed, *From Soul to Mind: The Emergence of Psychology from Erasmus Darwin to William James*.
5. See, for example, Thelen and Smith, *A Dynamic Systems Approach to the Development of Cognition and Action*; Port and van Gelder (eds), *Mind as Motion: Explorations in the Dynamics of Cognition*; Hendriks-Jansen, *Catching Ourselves in the Act*; Oyama, Griffiths and Gray (eds), *Cycles of Contingency: Developmental Systems and Evolution*.
6. E. O. Wilson, *Sociobiology: The New Synthesis*.
7. He characterises them, without citations, as the idea that mind is an obsolete fiction, '[like] the Tooth Fairy' (Pinker, *How the Mind*, pp. 77–8).
8. Ibid. p. 24.
9. Pinker might insist that flying presupposes, in addition to wings, a specific organ that solves the problem of aerodynamics, just as vision, according to the computational model of cognition, presupposes, in addition to eyes, a special mental organ that solves the problems of optics. But that would just move us to – and itself beg – the question of the validity of that model of cognition for birds, humans or anything else. The question is taken up below.
10. Barkow et al., 'Introduction', p. 8.
11. Ibid.
12. The drawing of distinctions between, on the one hand, neurophysiological structures and processes and, on the other, the 'functional properties' of the mind is a mark of what is called, in cognitive science, *functionalism* – that is, the idea, developed largely by artificial-intelligence theorists and rationalist philosophers of mind, that the operations that define intelligence (or reasoning) do not depend on their embodiment in any particular physical medium and could just as well be silicon chips in a computer as neurons in a living organism.
13. For general descriptions of such work in neuroscience, see Changeux, *Neuronal Man: The Biology of Mind*; Edelman, *Bright Air, Brilliant Fire: On the Matter of the Mind*; Damasio, *Descartes' Error.* For specific examples and technical description, see, for example, Ridderinkhof et al., 'The Role of the Medial Frontal Cortex in Cognitive Control'.
14. The method is referred to as 'Evolutionary Functional Analysis' in Tooby and Cosmides, 'Psychological Foundations'.
15. Pinker, *How the Mind*, pp. 21–2.
16. The account by Tooby and Cosmides is discussed in Section VI below.
17. Pinker, *How the Mind*, pp. 5–10.
18. See, for example, Maturana and Varela, *Autopoiesis and Cognition*; Thelen and Smith, *Dynamic Systems*; Christensen and Hooker, 'An Interactionist-Constructivist Approach to Intelligence'.
19. See, for example, Pinker, *How the Mind*, p. 57. Attempting to account for a persistent scepticism toward various assumptions of evolutionary psychology, Pinker writes of a 'takeover of humanities departments by the doctrines of postmodernism, poststructuralism, and deconstructionism' (ibid.). No citations are given for this absurd charge. The charge is repeated

with magnifications and elaborations, but still no citations, in Pinker, *Blank Slate*, p. 198.

20. See, for example, Clark, *Associative Engines: Connectionism, Concepts, and Representational Change*.
21. See, for example, van Gelder, 'What Might Cognition Be, If Not Computation?'; van Gelder, 'It's About Time: An Overview of the Dynamical Approach to Cognition'.
22. See, for example, Hutchins, *Cognition in the Wild*; Ingold, 'Technology, Language, Intelligence: A Reconsideration of Basic Concepts'; Hendriks-Jansen, *Catching Ourselves*.
23. Pinker, *How the Mind*, p. 22.
24. See, for example, Maturana and Varela, *Autopoiesis and Cognition*; Oyama, *Ontogeny of Information*; Bateson, 'Biological Approaches to the Study of Behavioral Development'; Christensen and Hooker, 'An Interactionist-Constructivist Approach'.
25. For an early and influential work on tacit, embodied knowledge, see Polyani, *Personal Knowledge*. For a recent analysis, informed by current research and theory in evolutionary biology, see Sheets-Johnstone, *The Roots of Thinking*.
26. See, for example, J. J. Gibson, *The Ecological Approach to Visual Perception*; Blake and Yuille (eds), *Active Vision*.
27. See Odling-Smee et al., 'Niche Construction'.
28. Pinker, *How the Mind*, p. 308.
29. Ibid.
30. Ibid. p. 57.
31. For accounts of category-formation by cognitive theorists who reject simple realist assumptions, see Lakoff, *Women, Fire, and Dangerous Things: What Categories Reveal about the Human Mind*; Griffiths, *What Emotions Really Are: The Problem of Psychological Categories*; Bowker and Star, *Sorting Things Out: Classification and its Consequences*.
32. Tooby and Cosmides, 'Psychological Foundations', p. 72.
33. For the history of the idea, see Rorty, *Philosophy and the Mirror of Nature*. Tooby and Cosmides refer to the notion of an evolved mind-world mesh not inaccurately as an 'evolutionary Kantian position' ('Psychological Foundations', pp. 69–70). The idea that humans are naturally selected to have objectively accurate perceptions of the world is also fundamental to some versions of so-called Evolutionary Epistemology. For discussion, see B. H. Smith, *Belief and Resistance*, pp. 49–51, 139–40, 167–8nn15–16.
34. Pinker, *How the Mind*, p. 355.
35. Pinker cites Lakoff and Johnson, *Metaphors We Live By*, to illustrate the existence of such concepts but omits mention of the latter's pointedly anti-innatist account of them as arising from universally recurrent bodily experiences.
36. The term 'mentalese' and much else in Pinker's account of concept-formation is adopted from the work of philosopher/cognitive theorist Jerry A. Fodor. Fodor's modularist-rationalist-innatist views are regarded sceptically by a number of other cognitive theorists (see, for example, Clark, *Associative Engines*; Hendriks-Jansen, *Catching Ourselves*; Christensen and Hooker, 'An Interactionist-Constructivist Approach').

37. Pinker, *How the Mind*, pp. 355–6.
38. Ibid. pp. 78.
39. Ibid.
40. For such objections, see, for example, R. C. Richardson, 'Evolution without History'; Sterelny and Griffiths, *Sex and Death: An Introduction to Philosophy of Biology*, pp. 313–36.
41. Pinker, *How the Mind*, p. 33.
42. For relevant critiques of innatist and/or genocentric accounts of such traits, see Deacon, *The Symbolic Species: The Co-Evolution of Language and the Brain*, pp. 334–43; Lieberman, *Eve Spoke: Human Language and Human Evolution*, pp. 125–32; Oyama, *Evolution's Eye: A Systems View of the Biology-Culture Divide*.
43. Pinker, *How the Mind*, p. 431 et passim.
44. According to Pinker, what is commonly referred to as 'learning' is actually the product of specific prewired mechanisms for acquiring particular behaviours (for example, language use). 'It [learning] is made possible by innate machinery designed to do the learning' (ibid., p. 33).
45. Pinker, *How the Mind*, p. 238.
46. 'Mixture', which suggests simple combination, should be something like dynamic interplay; 'nature and nurture', a popular cliché, might be better framed – more accurately as well as less 'tiresome[ly]' – as genetic and epigenetic processes.
47. Pinker, *How the Mind*, pp. 32–3.
48. It appears from Pinker's garbling of these matters that his knowledge of them is second-hand, derived mainly, his citations suggest, from hostile journalistic reports.
49. Pinker is especially exercised by the idea of 'cultural images', which, he argues (contra 'paranoi[d]' 'postmodernist and relativist' views), cannot shape our views of things or people (women, politicians and African-Americans are some of his examples) because images are 'linked' in the mind to 'a vast database of knowledge' of the things they 'stand for' (*Blank Slate,* pp. 215–16). Pinker implies – and seems to believe – that this interior encyclopedic knowledge (of, among other things, women, politicians and African-Americans) is objectively accurate and thereby duly corrective.
50. Pinker, *How the Mind*, p. 256.
51. See also Thornhill and Palmer, *A Natural History of Rape;* Buss, *The Dangerous Passion.*
52. Pinker, *How the Mind*, p. 467.
53. Pinker is either being obtuse here or taking advantage of the multiple meanings of 'sexuality' and 'biology' to foist an absurd claim on some of his critics. Although the idea of social construction can be invoked and applied crudely (as can any other idea, including natural selection), the point usually made by those citing it in these connections is not that people's physiological traits, erotic feelings or carnal activities ('human sexuality' in any of these senses) are independent of the evolutionary history of the species or of those people's individual genomes ('biology' in either of those senses). It is, rather, that conventional divisions, associations and normative attributions of various traits, feelings, activities and roles in accord with simple male/female dichotomies are not simple products or direct 'expressions' of either that

history or those genomes and that individual experiences and perceptions of such traits, feelings, activities and roles are not independent of those conventional divisions, associations and normative attributions.

54. Pinker, *How the Mind*, p. 467.
55. This supposed refutation of the idea that human sexuality (in whatever sense) could be socially shaped cleans the slate for Pinker's alternative account of its definitive determination (in presumably all senses) by the fitness-enhancing strategies of our mammalian and Upper Paleolithic ancestors. He concludes:

    > These conditions persisted through ninety-nine percent of our evolutionary history and have shaped our sexuality . . . A part of the male mind, then, should want a variety of sexual partners for the sheer sake of having a variety of sexual partners . . . Any bartender or grandmother would say . . .

    and so forth (*How the Mind*, pp. 468–9). The passage is discussed further in Chapter 7 below.
56. See note above. Pinker adds to their supposed testimony the results of a statistical survey of the dating and mating preferences of contemporary college men and women.
57. Tooby and Cosmides, 'Psychological Foundations', p. 39
58. Ibid., p. 118.
59. Ibid.
60. Tooby and Cosmides, 'Psychological Foundations', p. 118, italics in text. The idea of 'epidemiological' culture here owes to the notion of a parallel between cultural and biological evolution, with so-called 'memes' (units of cultural information) posited as counterparts to genes and mental 'infection' as the counterpart to genetic transmission (see Richard Dawkins, *The Selfish Gene*; Susan Blackmore, *The Meme Machine*).
61. Tooby and Cosmides, 'Psychological Foundations', p. 45.
62. Ibid., p. 48.
63. Ibid., p. 75.
64. Ibid.
65. The terms of Tooby and Cosmides' explanations of Evolutionary Functional Analysis are often misleading, with speculative reconstructions (for example, of Upper Paleolithic environmental conditions) and hypotheses (for example, of the existence of specific mental mechanisms) referred to as 'descriptions' and what should, accordingly, be modal terms (*if, would, were to*, and so on) replaced by quasi-observational terms (*when, did, does*, and so on). The final component of the method, referred to as a 'performance evaluation', is described as if it were an empirical test:

    > [It is] important . . . to see whether the proposed mechanism produces the behaviors one actually observes from [sic] real organisms under modern conditions. If it does, this suggests the research is converging on a correct description of the design of the mechanisms involved. ('Psychological Foundations', p. 74)

But, of course, one cannot 'see' whether a proposed mental mechanism produces some actual behaviour. One can only hypothesise that it does so on the basis of the same theoretical assumptions and selective observations that led one to propose it in the first place. Indeed, as a validation procedure, this component of the method appears to be a virtual prescription for self-affirming circularity.

For a classic analysis of the bootstrap logic involved in such accounts, see Gould and Lewontin, 'The Spandrels of San Marco and the Panglossian Paradigm: A Critique of the Adaptationist Programme'. What Gould and Lewontin characterise accordingly as 'just-so stories' are, Tooby and Cosmides reply drily, what, '[i]n science . . . is usually called explanation' ('Psychological Foundations', p. 77). The point of the Gould-Lewontin critique, however, is that the adaptationist scenarios in question, which depend on structures of individually dubious, mutually supporting assumptions, are tenuous or hollow explanations. R. C. Richardson, commenting on the disparity between such explanations and the relevant evidence in paleoanthropology, arrives at a similar conclusion: 'Our psychological capacities are surely the product of evolution. They may be adaptations. The explanations offered [by evolutionary psychology], though, are chimera' ('Evolution without History', p. 352).

66. Tooby and Cosmides, 'Psychological Foundations', p. 20.
67. Ibid., p. 21.
68. Ibid.
69. For comparably facile and tendentious historical accounts, see Wilson, *Consilience*, pp. 14–44; Pinker, *Blank Slate*, pp. 14–58.
70. See Lepenies, *Between Literature and Science: The Rise of Sociology*.
71. For a cosmopolitan exchange on the topic, see Changeux and Ricoeur, *What Makes Us Think?*.
72. See, for example, Carroll, *Literary Darwinism*, pp. ix–x.
73. For a sense of the range of objects of investigation, methods and theoretical perspectives, see the introductory essays in R. A. Wilson and Keil (eds), *The MIT Encyclopedia of Cognitive Science*, pp. xv–cxxxii.
74. For a proposal of (mutual) illumination along such lines, see Mallon and Stich, 'The Odd Couple: The Compatibility of Social Construction and Evolutionary Psychology'. As with many other two-culture bridging efforts, however (see, for example, Labinger and Collins, *The One Culture?*), the authors ignore the existence and effects of deeply – ideologically, institutionally and rhetorically – invested hostilities.

Chapter 7

Animal Relatives, Difficult Relations

The title of this chapter points to two sets of interrelated difficulties. Those in the first set arise chronically from our individual psychologically complex and often ambivalent relations to animals. The second set reflects the intellectually and ideologically crisscrossed connections among the various discourses currently concerned with those relations, including the movement for animal rights, ecological ethics, posthumanist theory, and such fields as primatology and evolutionary psychology. I begin with some general observations on kin and kinds – that is, relations and classifications – and then turn to the increasingly complex play of claims and counter-claims regarding the so-called species barrier.

I

The problem of our kinship to other animals mirrors that of our relation to other problematic beings: for example, the unborn, the mentally disabled, the drunk or the terminally comatose – beings, that is, who are recognisably our own kind but not yet, not quite, not just now, or no longer what we readily think of as *what we ourselves are*. In all these cases, there are difficulties handling both sameness and difference, difficulties framing the claims – either conceptual or ethical – of kinship, and, for formal philosophy, difficulties above all acknowledging just these difficulties.

Of course we are animals, it is said; or, to quote philosopher of ethics, Bernard Williams, 'The claim that we are animals is straightforwardly true,' the straightforwardness of the truth here deriving, it appears, from the current scheme of biological classification.[1] It is not always clear, however, that the classifications and distinctions of natural science – or, for that matter, vernacular ones either – should be awarded such unproblematic ontological authority.[2] When the issue is our responsibility to others, questions about limits are inevitably complicated by questions about *sorts*, and the relation between them broaches a domain we might call ethical taxonomy. Should we, for example, have care for dogs, cats, cows and horses but not birds, snakes or butterflies? For leopards and walruses but not lobsters or oysters? For all these, but not wasps, ticks or lice? Or for these, too, but not microbes or viruses?[3] Once the straightforward truth of our human distinctiveness is unsettled by the straightforward truth of our animal identity, there's no point, or at least no more obviously *natural* point, beyond which the claims of our kinship with other creatures – or, indeed, beings of any kind – could not be extended; nor, by the same token, is there any grouping of creatures, at least no more obviously *rational* grouping, to which such claims might not be confined.

My brief rehearsal, just above, of the chain of animate being was meant to evoke not only the variety of zoological kinds but also the disparateness of the domains in which we encounter them – on the streets and in our homes; on the farm and in the wild; at race tracks and circuses; in natural history museums and restaurants; beneath microscopes and in petri dishes. These juxtapositions are somewhat jarring, but this is to be expected. Each of these domains is likely to mark, for each of us, a specific history of experiential relations to the animal-kinds involved (as provisioners of, among other things, food, clothing, transportation, energy, company, creative inspiration and moral example; but also as parasites, predators and pathogens) and, with each such history, a repertoire of more or less specific attitudes and impulses. The impulses in question are deeply corporeal and, accordingly, when disturbed by sudden or dramatic domain-crossings (as in the juxtapositions above), likely to elicit that complex – jointly psychic and bodily – set of responses we call *cognitive dissonance*: that is, the sense of serious disorder or wrongness – and, with it, sensations of alarm, vertigo or revulsion – that we experience when deeply ingrained cognitive norms are unexpectedly violated. These responses are sometimes invoked by ethical theorists as our intuitive sense of outrage at what is thereby supposedly revealed as inherently improper or unjust: for example, the production of human embryos by cloning;[4] eating the flesh of dead animals; cutting open frogs, dogs or human beings for medical instruction; or the spectacle of two grown men in erotic embrace. This series is, of course, also somewhat jarring and, it might be objected, flagrantly indiscriminate. But I don't think so. For my point is not that these possibilities are all equally benign and acceptable or equally monstrous and unacceptable (I certainly don't see them that way myself), but that, with regard to the normative classification and treatment of other beings, it is hard to say where our individual judgements of impropriety and injustice start and stop being what we call rational or, put the other way around, where they start and stop reflecting the features of our individual histories and perhaps individual temperaments.

It is clear, I think, that all current conceptions and discourses of animals are marked by what one historian of taxonomy calls, in a slightly different connection, a 'polyphony' of classifications.[5] Problems arise because, in this domain of experience as elsewhere, categories are not abstract, neutral, inert containers but shifting tendencies to perceive and respond in some ways rather than others. Thus, in distinguishing a being as 'wild beast', 'domestic pet', 'livestock' or 'fellow-creature', we tap into a set of attitudes and expectations that are also bodily inclinations – for example, to approach or flee, capture or rescue, eat or feed it. These inclinations, complex enough in themselves, are also involved in our individual categorical norms: that is, in our sense of what, given a being of some kind, is the proper (natural, fitting) or improper (absurd, morally repulsive) way to feel about and deal with it. The significant variability of such norms is reflected in the cultural diversity of animal classifications and related practices.[6] As anthropologists never tire of reminding us, what one group eats, another worships (or worships *and* eats) and so forth. It is also reflected in the continuous possibility of the *re*classification of organisms and other beings and, accordingly, the transformation of related norms, attitudes and practices – a possibility that may serve as either (assuming we speak here of two things) strategy of indoctrination or instrument of enlightenment. Thus, foetuses may be cast as children, children as vermin, vermin as food-sources and food-sources as fellow-beings.

This last point can be spelled out a bit further. Just as opponents of abortion see, and strive to make others see, foetuses as babies and abortion, accordingly, as infanticide, so animal-rights advocates see, and strive to make others see, cows, rabbits, mice, monkeys, rats and seals as suffering fellow-creatures and, accordingly, the hunting, caging, killing, selling, wearing, riding or eating of them as oppression, murder, enslavement, exploitation or sacrifice. At the polemical centre of both movements are efforts to realign familiar classifications or effect analogous new ones and to draw, thereby, on previously established intuitions of propriety, rightness and wrongness. In both cases, these efforts depend as well on widely affirmed or assumed principles of ethical parity: for example, *treat like things alike* (thus protect infants from harm, whether born or unborn, human or fin-footed). Conversely, the counterpolemics of feminists, animal farmers and scientists defending, respectively, abortion, meat-eating or the use of mice and rabbits in research, consist largely of efforts to restore familiar distinctions (or reinforce alternative classifications) and to evoke, thereby, more favourable repertoires of intuitions – supplemented once again by what appear to be relevant principles of ethical parity: for example, *differences make a difference*; *do not treat as equal what is unequal*.

How the relevant cognitive/ethical norms and intuitions are formed, stabilised and transformed is a matter of some interest here, though also a matter of contention among contemporary anthropologists, psychologists and philosophers of mind.[7] Individual histories of interaction and particular cultural practices, including linguistic ones, are, of course, involved, but so also, it appears, are certain evolved, endemic tendencies: for example, a tendency to respond differentially to creatures with frontal versus dorsal eye-placement or to creatures that move bipedally rather than slither, scurry, swim or fly. Insofar as such tendencies reflect the evolutionary history of our own species, including the sorts of creatures with which our animal ancestors interacted (for better or worse, practically measured), some of our most profoundly intuitive responses to other animals, in this regard as others, reflect (for better or worse, ethically measured) our own animality.

Given the multiplicity and variability of the repertoires of responses we build up with respect to the relevant categories (animal, human, mammal, primate, beast, brute, living being and so forth), it seems inevitable that there will be clashes and conflicts within and between us in our ideas of propriety, naturalness, fitness and justice, and, conversely, of what constitutes absurdity, cruelty, inhumanity or injustice in our attitudes towards and treatment of other animals. The question is whether efforts to resolve such conflicts by appeals to putatively objective categories, rational distinctions or universal norms can avoid perpetuating, in *their* operations, the sorts of conceptual and social violence familiar from comparable axiological efforts in other spheres.[8]

Two further, related points may be added here. First, among the most extensively documented sites of continuity between humans and other animal species is that of *sociality* itself, including the ability to distinguish family members from non-kin and members of one's own social group from strangers, newcomers and outsiders. In responding strongly to members of certain animal species (for example, mammals) as kin or kind and, conversely, to members of other species (for example, snakes, insects and other invertebrates) as alien or remote, we exhibit capacities and rehearse impulses that are, in some of their origins and operations, extremely primitive.

Second, the imaginative intimacy of human with animal in myth, totem, fable and fantasy is no less profound in origin or, I think, significant in effect than the forms of kinship indicated by the observations of ethology or deductions of moral theory. Certainly the sources of our concepts of and responses to animals are not confined to what we might think of as our actual, empirical encounters with them. Thus, phoenix and unicorn no less than parrot or impala find quarter in the psychic bestiary, which has also been furnished, especially since Darwin and Freud, by an extensive literary phenomenology of animals. One thinks here of the vivid animal evocations of Hopkins and Rilke, Lawrence and Hemingway, Faulkner and Moore.[9] A recurrent topos among these and other (largely Modernist) writers is what could be called *the ontological thrill* of the animal: that is, the sense of a sudden intensification – quickening or thickening – of Being, as experienced, for example, at the sighting of a large bird or animal (hawk, deer, bear or snake) in the wild. Comparable sensations attend the hunting and indeed (or especially) killing of animals, as well as riding them, wearing their skins or consuming them as food, and are also involved in fantasies of coupling with, being or becoming them.[10] It would not be a simple matter, I think, to disentangle these primitive sensations and animistic identifications from the impulses that constitute our most intellectually subtle and ethically potent intuitions of animals or, thereby, our most reflective and respectful relations with them.

II

I turn now to the intellectual terrain on which these psychologically complex and often emotionally and ethically ambivalent relations to animals are currently played out, focusing here on the issue of the continuity or discontinuity between humans and other species.

To begin at a relatively simple entry point, there is, of course, the argument for continuity from shared DNA – 98.5 per cent, by the latest count, in the case of humans and chimpanzees – and also recent fieldwork in primatology: Jane Goodall's observations of tool use among apes; Sue Savage-Rumbaugh's accounts of the evidently spontaneous acquisition of language by bonobo chimps; Frans de Waal's studies of social and arguably proto-ethical behaviour (food-sharing, peace-making and so forth) in various primates; and reports by these and other ethologists of the non-genetic and arguably proto-cultural transmission of skills and information among members of other species.[11] The tendency of all these studies – and of others that examine the complexity of the emergence of many so-called instinctive behaviours in birds and other animals (birdsong, migration-patterns and so forth)[12] – is to challenge or at least complicate classic humanistic accounts of the crucial difference between humans and other species.

It must be added, however, that weighty as the DNA figure is, the species barrier, as biologically defined, appears to hold.[13] That is, bestiality in the sexual sense, however fertile in myth or dream, has no documented issue. To be sure, the possibility, now as ever, haunts the imagination, at least the *human* imagination (who knows the dreams of dogs or sheep?): for example, in Greek myth, where access to godhead is mediated by union with animals (or perhaps it is the other way around),[14] or in John Frankenheimer's film of H. G. Wells's story, *The Island of Dr Moreau*, where a union of moralised Darwinian fantasy and late-twentieth-century visual technology issues in some highly engaging, though ultimately melancholy, progeny.[15] Nevertheless, it seems to be the case that man-beast relations are not literally reproductive.

Moreover, work by other primatologists and ethologists – or, indeed, the same ones – casts doubt on a number of familiar assumptions regarding the identity, continuity or even just comparability, of various human and animal capacities. For example, Terrence Deacon, a biological anthropologist and brain researcher with no apparent professional or ideological investments in an insuperable species barrier, makes a good case for the reciprocally selective co-evolution of key features of (1) human sociality and communication and (2) the increasingly distinctive size, structure and operations of the human brain and, accordingly, for the claim that symbolic communication (duly defined and explained) and related social and cognitive skills emerge reliably only in human communities – human *communities*, not human *beings*, which honours the bonobos' achievements even as it helps account for their rarity.[16] Similarly, Michael Tomasello, a developmental psychologist who has collaborated with Savage-Rumbaugh, documents subtle but developmentally crucial differences in certain types of behaviour in apes and human children that are generally taken to be the same in both (for example, so-called imitative behaviour) and that have led other psychologists to the dubious attribution of human-like capacities (for example, intentional deception) to apes.[17]

These studies do not, of course, cancel each other out – not, that is, unless one is keeping very crude tallies ('here's one for the chimps, there's one for the humans' and so on). They do indicate, however, that, with respect to the sorts of capacities commonly invoked in these debates (language, culture, social learning, a moral sense, rationality, deception and so forth), the question of the continuity of humans with other species cannot be posed as a simple alternative or even as a simple matter of degree. In some ways, by some calculations, with regard to some traits, the permeability of the species barrier seems increasingly manifest; in other ways, by other measures, with regard to other tendencies and capacities, significant disjunctions between humans and other animals are being documented and incorporated into biological and behavioural theory. Nor, for the same reasons, can the ethical issues raised by animal rights advocates or posthumanist theory be decided by current findings in genetics or ethology. There are too many dimensions of potential identity and/or distinctiveness and, of more fundamental significance, there is no way of assessing their relative importance that does not risk begging the very questions that such empirical findings are supposed to resolve.

An issue of particular interest here is which species do and do not possess (or exhibit) 'culture', controversies over which illustrate and exacerbate the chronically perplexed relations between empirical science and rationalist/humanist moral theory. Thus posthumanist Cary Wolfe, though cautioning against 'naturalism in ethics', cites Goodall's observations of chimpanzee tool-use to challenge the claim by humanist Luc Ferry of a uniquely human capacity for culture,[18] while humanist Alan Wolfe challenges naturalist J. T. Bonner's claim that culture emerged with pre-human primates by noting that the argument depends on a dubiously ad hoc and otherwise irrelevant definition of culture as any non-genetic transmission of behaviour.[19] At the same time, we recall, evolutionary psychologists John Tooby and Leda Cosmides deny the existence of culture so defined – that is, as the non-genetic transmission of behaviour – among humans or any other species because, in their view, *all* significant transmission of behaviour is genetically based.[20] I return below to the problematic claims of evolutionary psychology in these regards and to the difficult relations indicated here between classic humanism and various post-, anti-, and non-humanisms. First, however, we should take note of the perplexed issue of animal 'minds'.

Since the beginning of the twentieth century and especially with the ascendancy of positivism in psychology and behaviourism in the study of animals, claims that animals can think, have consciousness, or are self-aware have elicited routine charges of 'anthropomorphism', meaning, in these instances, the gratuitous attribution to members of other species of so-called higher mental processes (reasoning, deliberation, calculation and so forth) to account for behaviours that could be explained (or so the charge implies) by simpler mechanisms – for example, by instincts, conditioned reflexes, rote or trial-and-error learning, unconscious prompting by trainers or the physical effects of chemical traces.[21] In recent years, however, claims about the capacities of animals for thinking and for relatively complex forms of feeling and intentionality (guilt, blame, remorse, self-sacrifice, deception, revenge and so forth; the nice technical term here is 'anthropo*path*ism') have had a more receptive hearing, especially in fields such as cognitive psychology that have developed in close association with traditional rationalist/intentionalist/representationalist philosophy of mind.[22] Thus a recent article argues for the propriety of ascribing higher mental processes, specifically as described by such philosophers of mind as John Searle and Roger Scruton, to at least the sorts of animals with which (or, perhaps, with *whom*) we live and work.[23]

Such arguments, however, can backfire in curious ways; for, in lowering the species barrier, we open a two-way street. Thus the claim of continuity between animals and humans with regard to higher mental processes makes it possible to ask whether comparable charges of 'anthropomorphism' (in the sense given above) could not be levelled against traditional rationalist interpretations of *human* actions – which is, of course, what behaviourism always maintained. The plausibility of such reverse charges (that is, that explanations of human behaviour in terms of deliberations, interior representations and so forth introduce gratuitously rational processes and mentalistic concepts) is strengthened by recent work in robotics, where relatively simple mobile machines 'learn' to negotiate their environments successfully without pre-programming, calculations or interior representations of any kind and, in so doing, exhibit what appear to be purposive and quite human- or animal- (or at least insect-) like behaviours.[24] There appears, in other words, no clear tendency in contemporary cognitive research to validate the claims of one side over the other in the debate between continuists and discontinuists with regard to 'higher mental process', and a victory via one demonstration may be overturned in another or turned into a defeat from another perspective.

Contrary to some animal rights advocates, I do not believe that parity of reasoning forbids any difference in the ways we interpret the behaviour of humans and that of other animals, just as I do not believe that ethical parity requires uniformity in our practical treatment of each. I would suggest, however, that once we admit the propriety of anthropomorphising animals at least some of the time, it becomes harder to see why we should not accept the propriety of naturalising humans – at least some of the time. The latter possibility, as reflected in the thought of, among others, Machiavelli, La Mettrie, Spinoza, Nietzsche, Freud and Skinner, has been recurrently denounced as reductive, cynical, and/or sinister and strenuous resistance to it continues not only to be pressed in humanistic thought but virtually to define it. Such reactions recur in current responses by philosophers and other academics and intellectuals, explicitly humanistic and otherwise, to the claims of sociobiology or evolutionary psychology, but often, as shall be seen below, in quite ideologically tangled ways.

III

In his book *How the Mind Works* (discussed in Chapter 6), evolutionary psychologist Steven Pinker introduces a resolutely naturalistic account of human sexual relations with a double negative that sharpens and secures his wanted emphasis: 'The human mating system is not like any other animal's. But that does not mean it escapes the laws governing mating systems, which have been documented in hundreds of species.'[25] Reinforcing the emphasis through a determined fusion of idioms, Pinker continues:

> Any gene predisposing a male to be cuckolded, or a female to receive less paternal help than her neighbors, would quickly be tossed from the gene pool. Any gene that allowed a male to impregnate all the females, or a female to bear the most indulged offspring of the best male, would quickly take over.[26]

Here, as elsewhere in the popular sociobiological literature, dubious subsumptions of the human by the pan-zoological are mediated by the casual conjunction of, on the one hand, technical and ostensibly generic (that is, non-species-specific) terms such as 'male', 'female', and 'gene pool' with, on the other hand, vernacular terms such as 'cuckolded', 'neighbors', and 'indulged offspring' that evoke familiar human situations and attitudes. Thus Pinker's account continues:

> What kind of animal is *Homo sapiens*? We are mammals, so a woman's minimal parental investment is much larger than a man's. She contributes nine months of pregnancy and (in a natural [sic] environment) two to four years of nursing. He contributes a few minutes of sex and a teaspoon [poignant human detail] of semen . . . These facts of life have never changed . . . A part of the male mind . . . should want a variety of sexual partners . . . Any bartender or grandmother you ask would say . . .[27]

and so forth.

It is, of course, just such accounts that evoke from traditional humanists the most energetic affirmations of a crucial difference between humans and other animals. I will turn to these shortly but would note, first, a significant element in the theoretical framework of evolutionary psychology that complicates its intellectual profile and distinguishes it from first-generation sociobiology, namely, its supplementing the latter's familiar adaptationist accounts of human behaviour with the more recently developed computational model of mind. The central idea here, we recall, is that the human mind, like an artifactual computer, is an information-processing device 'engineered' (in this case by natural selection, represented as a quasi-purposive agent) to solve environmentally posed problems by performing operations on symbols in accord with 'hardwired' (in this case innate, species-wide, genetically transmitted) rules. As indicated in Chapter 6, the computational model of mind and its specific appropriations in evolutionary psychology are seen as problematic by many theorists in the relevant fields, that is, evolutionary biology, genetics, neuroscience, computational theory and artificial intelligence.[28] What makes that model of mind significant in the present context, however, is that, in invoking and promoting it, evolutionary psychologists are led to stress the *distinctiveness* of human cognitive capacities in contrast to those of other animals.

Thus we find the following in Pinker:

> Some authors are militant that humans are barely different from chimpanzees and that any focus on specifically human talents is arrogant chauvinism or tantamount to creationism . . .[29]

> We *are* naked . . . apes that speak, but we also have minds that differ considerably from those of apes. The outsize brain of *Homo sapiens* is, by any standard, an extraordinary adaptation. It has allowed us to inhabit every ecosystem on earth, reshape the planet, walk on the moon, and discover the secrets of the physical universe. Chimpanzees, for all their vaunted intelligence, are a threatened species clinging to a few patches of forest and living as they did millions of years ago . . . We should not be surprised to discover impressive new cognitive abilities in humans, language being just the most obvious one.[30]

Pinker's catalogue of distinctive human achievements, which omits mention of any artistic, philosophical or – aside from technology – other cultural accomplishments, will strike many readers as distinctly lopsided as well as otherwise distasteful.[31] The idea that the highest reaches of primate intelligence are exhibited in feats of ecological expansion and technology ('inhabit every ecosystem . . . reshape the planet, walk on the moon' and so forth) is altogether consistent, however, with a conception of the human mind as a computational mechanism designed to solve environmentally posed problems *and* with a conception of art, philosophy and other cultural activities as, at best, secondary and superficial elements of the conduct and conditions of human life. What is especially notable here is that the definitive conjunction of these two ideas in evolutionary psychology yields a view of human beings as *absolutely continuous* with other animals with respect to social and sexual behaviour and *absolutely discontinuous* from them with respect to cognition. To judge from the current vogue of evolutionary psychology among members of the educated public and the ever-increasing number of popular publications applying its analyses to an ever-widening range of human practices from rape and infanticide to landscape painting and Romantic poetry,[32] this is a settlement of the question of our kinship with other animals (and a characterisation of humanity – that is, like beasts in our social and sexual behaviour and like computers in how we think) that currently appeals to a great many people.

But, of course, it does not appeal to everyone. On the contrary, explanations of human actions, motives, emotions and mental life in terms of mammalian reproductive strategies and mechanical computations appear absurdly shallow, callow, oversimplified or irrelevant to audiences of many sorts and elicit from traditional humanists renewed affirmations of a crucial difference between humans and both animals and machines: for example, Alan Wolfe's *The Human Difference: Animals, Computers, and the Necessity of Social Science*. Significantly, however, Wolfe's strenuous defence of a uniquely human domain of being – and, with it, a special domain of knowledge – is directed not only against the naturalising, mechanising claims of sociobiology and artificial intelligence *and* the species-egalitarian arguments of animal rights advocates, but also against what he sees as the 'nihilistic' anti-humanism of 'postmodern' thought. 'Postmodernism', along with Deep Ecology, ecofeminism, and the movement for animal rights, are also the featured targets of French philosopher Luc Ferry, who argues that all these developments are regressive repudiations of the classic humanistic principles and related emancipatory ideals of the French Enlightenment. 'Postmodernists' are also central objects of scorn in the work of sociobiologists and evolutionary psychologists themselves, but here in company with 'humanists' per se (along with feminists and emancipatory-minded persons of any sort) and in contrast to what are represented as duly fact-knowing, fact-facing scientists.[33]

Clearly the issue of our kinship with animals produces strange bedfellows, joining those commonly and otherwise adversaries, setting at odds those commonly and otherwise allies, and revealing intellectual commitments, tastes and aversions that are evidently powerful but otherwise obscured. An especially bemusing reconfiguration of this kind appears in the conceptual connections and, in some cases, shared political orientation between, on the one hand, *ecological ethicism*, as embodied in the animal rights movement, ecofeminism, radical environmentalism, and ecology-minded moral theory, and, on the other hand, *scientific and philosophical naturalism*, as embodied in, among other projects, sociobiology. What both general positions have in common is the idea of a single natural order, which, in the case of ecological ethics, joins human beings with all other forms of life and ultimately 'all of nature' and, in the case of philosophical or scientific naturalism, unites us with all other organisms, ultimately all physical phenomena, and thus again, but in a different sense and from a different perspective, 'all of nature'.

For ecological ethics, this ontological unity implies a *moral* imperative to extend our regard for the welfare of fellow human beings to other animals, other living beings, and, in some arguments, to all other *beings*, living and sentient or otherwise. Thus John Llewelyn argues, by way of Martin Heidegger and Emmanuel Levinas, that we have moral responsibilities to all things that 'have need of us'.[34] The major concrete examples of such things offered by Llewelyn are trees cut down for paper used in pulp magazines of questionable intellectual value, mountains about to be turned into ski slopes by enterprising developers, and large rocks outside Edinburgh (where he is Professor of Philosophy) defaced by, it appears, local vandals. Llewelyn acknowledges, but does not address, the piquant problems presented by any attempt to work out this moral imperative in terms of specific actions or policies, including policies with regard to human beings of apparently different classes and tastes.

Correspondingly for scientific and philosophical naturalism, the posited fundamental unity of nature implies an *epistemic* imperative to integrate accounts of human behaviour with current scientific understandings of all other *natural phenomena*, from the behaviour of animals to that of quantum particles. If, as just suggested, the moral imperatives of ecological ethics can appear logically, practically or (indeed) ethically dubious, the epistemic imperatives of naturalism can certainly translate, as in Pinker's case, into a vulgar eagerness to transfer explanations as rawly as possible from barnyard and jungle to contemporary human societies. The difficulties indicated here are not, in my view, intrinsic either to a strong concern for the natural environment and/or respect for life or to a conviction of the intellectual and perhaps ethical interest of naturalised accounts of human behaviour. They may be intrinsic, however, to any ethical or epistemic project insofar as it conceives itself in all-trumping – exclusivist, supremacist, and/or universalist – terms.

Some of the ideological and political perplexities that emerge when the affinities and divergences indicated above are played out institutionally are reflected in John Dupré's *The Disorder of Things: Metaphysical Foundations of the Disunity of Science*, already encountered in Chapter 4. Dupré, a philosopher of biology, laments what he sees in his field as a 'baleful . . . reverence for the products of science verging often on the obsequious',[35] as exemplified by 'the current move to "naturalize" epistemology by appeals to so-called cognitive science; the project of replacing philosophy of mind and apparently the mental with neurobiology . . . and in ethics, the idea that the speculations of sociobiologists might tell us something about how we should live'.[36] Dupré's defence, accordingly, of a crucial distinction between science and philosophy echoes the defence by Alan Wolfe of a crucial difference and proper disciplinary barrier between the natural and the social (or 'human') sciences; and both defences echo, assume and appeal to the species barrier – that is, to the classic humanistic idea of an essential distinction between humans and other animals. Conversely but also quite consistently, E. O. Wilson, founder of sociobiology (though also, as it happens, a strenuous environmental activist),[37] invokes the idea of a single natural order to argue for a dissolution of all disciplinary barriers and an integration of existing fields of knowledge along lines that would put Dupré, Alan Wolfe, Cary Wolfe, John Llewelyn, Bernard Williams and many readers of this book out of business – that is, as the subsumption and absorption of the social sciences, philosophy and the other humanities by a single, comprehensive natural-science discipline,[38] and, thereby, a determined repudiation of what Wilson sees as the obsolete and essentially theological distinction between humans and other animals.[39]

IV

Well . . . where does that leave us? Who is friend, who enemy here? Which distinctions do we wish to preserve and which to see dissolved? And are we sure, in all this, that we know – or agree – who 'we' are?

Such questions cannot be answered simply or finally and, in a sense, cannot be answered at all. Rather, they restate the fundamental difficulties involved in any attempt to determine in a formally principled or univocal way – whether scientific or philosophical, naturalistic or humanistic – our relations to other creatures. This is not, in my view, a despairing observation. On the contrary, what it indicates is the necessary openness of these questions to ongoing address. When all the arithmetic is a priori and the conclusions all foregone, there is no intellectual or ethical activity at all, just the animation (so to speak) of a set of mechanical (so to speak) procedures.[40] In operating without fixed or formal principles, one is confronted, of course, with the need for continuous attention and responsiveness – for investigating and registering, remembering and imagining, comparing and assessing, and deciding under conditions of radical uncertainty and, in a sense, incoherence.[41] These requirements, however, could be seen to constitute the very activity of responsible reflection, to define the very conditions of what we – some of us, anyway – call ethical judgement and action.

A final word, accordingly, on animal ethics. Because of the types of sociality that human beings share with most other primates, our attitudes and behaviour toward members of our own and other species – including our intuitive sense of what (or who) can and cannot be sold, beaten, killed or eaten – are shaped and sustained by, among other things, the example and approval or disapproval of other people, at least those we see as our own kind. The distaste that some of us have come to feel for such once well-established practices as eating meat or casually destroying animals in scientific experimentation is, I believe, as much the product of social example and sanctions of these kinds as of critical reflection per se. But so also was our previous taste or tolerance for such now rejected practices. In other words, our specifically ethical impulses and attitudes toward animals seem to draw force, for better or for worse, from aspects of our specific hominid nature, just as our more general impulses, attitudes and practices seem to draw force – again, for better or for worse – from aspects of our more general animal nature.

For better or for worse: that necessarily ambivalent, contingent assessment has been a key point all along. It is clear, I think, that our biological kinship with other animals shapes and shadows some of our most compelling appetites and anxieties, exalting and destructive impulses. The recognition of and response to our own animality in these respects is an old story, told many times over in fable and sermon and still being told in discourses from psychoanalysis to neurophysiology. We have certainly not, as a species, society or civilisation, transcended or determined that we want to or *could* transcend our psychic and corporeal kinship to other creatures. Nor have we resolved the conceptual and ethical problems presented by that complex, ambivalent connection: neither in the established traditions of Western philosophy nor, it appears, in even the most pertinent and elaborated reflections of non-Western traditions. It may be that the most distinctive contribution of contemporary (or 'postmodern') thought in this regard has been to allow the inevitability and power of these ambivalences to be acknowledged as such and also to allow – or to insist on – their entry into whatever ethics we devise. As 'posthumanists', we have begun to chart the costs and limits of the classic effort to maintain an essential species barrier and have sought to diminish those costs and to press against those limits in our own conceptual and other practices. The *telos* – aim or endpoint – of these developments is conceived here, however, not as the universal recognition of a single, comprehensive order of Nature or Being but, rather, as an increasingly rich and operative appreciation of our irreducibly multiple and variable, complexly valenced, infinitely reconfigurable relations with other animals, including each other.

Notes

1. B. Williams, 'Prologue: Making Sense of Humanity', p. 15. Williams observes, 'We are a kind of animal in the same way that any other species is a kind of animal – we are, for instance, a kind of primate' (p. 13).
2. On the problematic status of the species concept in contemporary biology, see Mayr, *Toward a New Philosophy of Biology*, pp. 315–34; and Ereshefsky, 'Species Pluralism and Anti-Realism', pp. 103–20. Williams's central concern in the essay is not to establish the ontology of human beings as such but, having acknowledged the biological classification, to affirm the values of a cultivated humanism and the associated projects of analytic/rationalist philosophy in the face of such arguably rival enterprises as cognitive science, cultural studies and posthumanist ethics. The essay concludes with a defence of 'a humanistic account of the human species', distrust of which Williams identifies with 'a distrust of . . . despair at . . . or hatred of' the 'quality' of 'humanity' ('Prologue', p. 22).
3. Philosopher and animal rights advocate Tom Regan cites evolutionary theory as his authority for drawing an ethically freighted line between 'mammalian animals' and all other beings (Regan, *The Case for Animal Rights*, p. 403n1). Given, however, his explicit criterion for moral status, namely, being 'the-subject-of-a-life', it is hard to see what warrant he could find in contemporary evolutionary theory for drawing the line at just that point.
4. Current public opposition to such a procedure is described by Gilbert C. Meilaender, Professor of Christian Ethics at Valparaiso University and member of a national panel of bioethical advisors, as 'a natural repulsion' (qtd. in Stolberg, 'Bush's Advisers on Ethics Discuss Human Cloning').
5. See Ritvo, *The Platypus and the Mermaid, and Other Figments of the Classifying Imagination*. One recalls that Borges's parody of systematic knowledge features an animal taxonomy – 'On those remote pages it is written that animals are divided into (a) those that belong to the Emperor, (b) embalmed ones, (c) those that are trained, (d) suckling pigs, (e) mermaids, (f) fabulous ones, (g) stray dogs', and so forth (Borges, *Other Inquisitions*, p. 108) – and that the list was rehearsed to comparable ends by Foucault at the beginning of *The Order of Things*.
6. For examples and discussion of vernacular animal classifications in various cultures, see Lakoff, *Women, Fire, and Dangerous Things*, pp. 46–52.
7. See Lakoff, *Women, Fire, and Dangerous Things*. The notion of 'prototypes', originally developed by psychologist Eleanor Rosch and elaborated in Lakoff, is especially pertinent here. Experimental subjects report that certain members of a given category appear to be more representative than others. Thus lions are seen as more typical of 'animals' than vampire bats, robins as more truly 'birds' than ostriches, and so forth. Lakoff observes that the internal structure of such basic-level categories as *chair*, *animal* and *bird* apparently 'depend[s] not on objects themselves . . . but on the way

people interact with objects: the way they perceive them, image them, organize information about them, and behave toward them with their bodies' (ibid., pp. 50–1).

8. See B. H. Smith, *Contingencies of Value* and *Belief and Resistance*, for the problematic operations of formal axiology in, respectively, aesthetics and epistemology.
9. See, for example, William Faulkner, 'The Bear'; G. M. Hopkins, 'The Windhover'; D. H. Lawrence, 'Snake' and 'The Fox'; Marianne Moore, 'The Elephants'; and R. M. Rilke, 'The Panther'. One might also note here the multiple animal evocations of popular culture, from children's cartoons to corporate logos.
10. See Cixous, 'Love of the Wolf' and Deleuze and Guattari, 'Becoming-Animal'. For some striking images in contemporary visual art, see Baker, 'What Does Becoming-Animal Look Like?'.
11. Goodall, *The Chimpanzees of Gombe: Patterns of Behavior*; Savage-Rumbaugh et al., *Apes, Language, and the Human Mind*; de Waal, *Good Natured: The Origins of Right and Wrong in Humans and other Animals*; Cheney and Seyfarth, *How Monkeys See the World*.
12. See Bateson (ed.), *The Development and Integration of Behaviour: Essays in Honour of Robert Hinde*.
13. Reproductive isolation, along with shared physiological features and presumed common phylogenetic descent, remains a major criterion for identifying and distinguishing biological species, though contemporary taxonomists acknowledge that the groupings yielded by these characteristics do not always coincide (see Ereshefsky, 'Species Pluralism').
14. For a sophisticated contemporary interpretation, see Calasso, *The Marriage of Cadmus and Harmony*.
15. For other recent examples of imagined man-beast unions, see Baker, 'Sloughing the Human' and 'What Does Becoming-Animal Look Like?'; Høeg, *The Woman and the Ape*.
16. Deacon, *The Symbolic Species*. For an up-to-date account of the perennial idea of language as distinguishing the human from the animal, see C. Wolfe, *Animal Rites: American Culture, the Discourse of Species, and Posthumanist Theory*, pp. 44–94. For a strong critique of the idea and its legacy in contemporary linguistics, see Savage-Rumbaugh et al., *Apes*, pp. 75–138.
17. See Tomasello et al., 'Imitative Learning of Actions on Objects by Children, Chimpanzees and Enculturated Chimpanzees'. On the idea of dissimulation as the mark of the human, see Derrida, 'And Say the Animal Responded?'.
18. C. Wolfe, *Animal Rites*, pp. 21–43. Wolfe seeks to formulate an explicitly nonhumanist but also emphatically nonnaturalist and nonempiricist principle for the treatment of animals. 'Taking account of the ethical relevance of ethologists like Goodall,' he writes, '*does not mean committing ourselves to naturalism in ethics*' (p. 42; italics in text); nor, he continues, need we

> cling to any empiricist notion about what Goodall or anyone else has discovered about nonhuman animals . . . to insist that when our generally agreed-on markers for ethical consideration are observed in species other than *Homo sapiens*, we are obliged to take them into account equally and to respect them accordingly.

Wolfe explains his argument here as 'amount[ing] to nothing more than taking the humanist conceptualization of the problem at its word [that is, the idea that humans are uniquely ethical subjects because they possess certain presumptively distinctive features] and being rigorous about it', his aim being to show that humanism must 'generate its own deconstruction' when thus rigorously pursued (ibid.). It could be suggested, however, that 'being rigorous about it' is precisely what creates the greatest conceptual and ethical difficulties, whether 'it' (the subject of that rigour) is Ferry's humanism, Singer's utilitarianism (also examined by Wolfe, *Animal Rites*, pp. 33–6), or any other formal doctrine – including Wolfe's own posthumanism insofar as it frames itself as an absolute principle.

19. A. Wolfe, *The Human Difference: Animals, Computers, and the Necessity of Social Science*, pp. 36–40.
20. The concept of 'culture', they argue, is obsolete, being properly replaced by the idea of the individual's 'inferential reconstruction' of 'information' and 'rules' based on innate mental computational mechanisms (Tooby and Cosmides, 'Psychological Foundations', esp. pp. 117–23). See the discussion of Tooby and Cosmides in Chapter 6 above.
21. An early proponent of such claims – and object of such charges – was George Romanes. For a good account, see Leahey, *A History of Psychology: Main Currents in Psychological Thought*, pp. 251–3. The classic statement of principled anti-anthropomorphism was formulated in 1894 by comparative psychologist C. Lloyd Morgan. For discussion, see Sober, 'Morgan's Canon'.
22. See esp. Griffin, *Animal Thinking*. For an overview of more recent developments, see Ristau, 'Cognitive Ethology'. For a biologically informed analysis of the general issue, including a critique of the tendency among cognitive scientists and philosophers of mind to affirm a rigorous biological naturalism while maintaining a sharp distinction between the cognitive operations of humans and those of other organisms, see Sheets-Johnstone, 'Consciousness: A Natural History'.
23. See Cox and Ashford, 'Riddle Me This: The Craft and Concept of Animal Mind'.
24. See Brooks, 'Intelligence without Representation'.
25. Pinker, *How the Mind Works*, p. 467.
26. Ibid.
27. Ibid. pp. 468–9.
28. See Chapter 6 above, esp. Section II.
29. Pinker's allusion here (he gives no citations) is most obviously to animal rights advocates but, as the context makes clear, also to scientists and philosophers who dispute the arguably neo-creationist idea of human mental exceptionalism – for example, Sheets-Johnstone, 'Consciousness'; Savage-Rumbaugh et al., *Apes*; Ristau, 'Cognitive Ethology'.
30. Pinker, *How the Mind*, pp. 40–1. Aspects of this statement may recall Deacon's theory (mentioned above) of the co-evolution of human communication, cognitive social skills and brain structure. There is, however, a crucial difference. Whereas Pinker sees these skills as prewired and innate, Deacon stresses that their emergence and development require structured social interactions in human communities. For discussion of the difference, see Deacon, *The Symbolic Species*, pp. 140–2, 328–33.

31. The terms of Pinker's contrast between humans and chimpanzees duplicate fairly conspicuously the sorts of racial, sexual and/or ethnic self-glorifications used to justify various forms of political and social discrimination among humans, for example, by men against women or by Europeans against Asians and Africans. The similarities between the two are routinely invoked by animal rights advocates (see, for example, Jamieson, *Morality's Progress: Essays on Humans, Other Animals, and the Rest of Nature*) and posthumanists (see, for example, C. Wolfe, *Animal Rites*, p. 7) as evidence of an ethically significant slide between speciesism and both racism and sexism and also as an argument for the historical inevitability of a general acknowledgement of the rights of animals. Since Pinker is scornful of what he calls 'fashionable "liberation" ideologies' (*How the Mind*, p. 48), his duplication of the terms of those self-glorifications here may be intentionally abrasive.
32. See, for example, Barkow, *Darwin, Sex and Status: Biological Approaches to Mind and Culture*; Wright, *The Moral Animal: The New Science of Evolutionary Psychology*; Thornhill and Palmer, *A Natural History of Rape*; Buss, *The Evolution of Desire: Strategies of Human Mating*; A. Richardson and Steen (eds), *Literature and the Cognitive Revolution*; Easterlin (ed.), *Symposium: Evolution and Literature*.
33. See, for example, Pinker, *How the Mind*, pp. 48, 57, 492–3; E. O. Wilson, *Consilience*, pp. 40–4.
34. Llewelyn, *The Middle Voice of Ecological Conscience; A Chiasmatic Reading of Responsibility in the Neighborhood of Levinas, Heidegger, and Others*, pp. 245–77.
35. Dupré, *The Disorder of Things*, p. 13.
36. Ibid., p. 268n12.
37. See E. O. Wilson, *The Diversity of Life* and (ed.) *Biodiversity*.
38. E. O. Wilson, *Consilience*, pp. 181–265. Such unification projects are discussed in Chapters 5 and 6 above. It is not surprising that the title of Wilson's book reverses almost exactly that of Dupré's: the conflict between their views could hardly be starker. For Dupré's elaboration of his objections to sociobiology and evolutionary psychology, see his *Human Nature and the Limits of Science*.
39. The insistent (and indeed born-again) anti-theological element of Wilson's scientism (science, he tells readers, replaced his native Southern Baptism [*Consilience*, p. 6]) is worth noting. Like a number of other environmental activists, he maintains that the natural world can be protected from the present mindless forces of depredation and depletion only through a powerful counterforce of enlightenment and, with it, a revolutionary transformation of social, cultural and political priorities and practices. There is, however, a crucial divergence of vision here. For Wilson, the required enlightenment would be a return to and fulfilment of Enlightenment ideals, notably (as he sees them) the authority of reason and science in opposition to irrationalism and religious superstition, and would issue in, among other things, a thorough *naturalisation* of the universe. For many other environmental activists, however (for example, Berry in *The Unsettling of America: Culture and Agriculture* and Abram in *The Spell of the Sensuous: Perception and Language in a More-than-Human World)*, the required enlightenment

would be a repudiation of just those Enlightenment ideals (seen here as leading to and justifying the exploitation and destruction of nature) and would issue in a thorough *sacralisation* of the universe.

40. In connection with the question of justice toward animals (among other issues), Jacques Derrida remarks our sense that

> each case is other, each decision is different and requires an absolutely unique interpretation, which no existing, coded rule can or ought to guarantee absolutely. At least, if the rule guarantees it in no uncertain terms, so that the judge is a calculating machine – which happens – we will not say that he is just, free and responsible. (Derrida, 'Force of Law: The "Mystical Foundation of Authority" ', p. 961)

He remarks, similarly, our sense that '[a] decision that didn't go through the ordeal of the undecidable would not be a free decision, it would only be the programmable application or unfolding of a calculable process' (ibid., p. 963). With an eye, however, on the aporias of common intuitions and conceptions of justice, Derrida goes on to observe that 'we also won't say [the decision is just] . . . if [the judge] improvises and leaves aside all rules, all principles' (ibid., p. 961) and that any decision, once made, can be seen to have followed, invented, reinvented or reaffirmed some rule (ibid., p. 963). For related reflections on the ethics of animal-human relations, see Derrida, ' "Eating Well" or the Calculation of the Subject'.

41. On what could be seen as the irreducible incoherence of our individual actions and our practices of conceptualisation and judgement, see B. H. Smith, *Contingencies*, pp. 147–9.

Works Cited

Abbott, Andrew (2001), *Chaos of Disciplines*, Chicago: University of Chicago Press.

Abram, David (1997), *The Spell of the Sensuous: Perception and Language in a More-than-Human World*, New York: Vintage.

Alcoff, Linda Martin (1995–6), 'Is the Feminist Critique of Reason Rational?', *Philosophic Exchange* 26, pp. 59–79.

Appleby, Joyce, Lynn Hunt and Margaret Jacob (1994), *Telling the Truth about History*, New York: Norton.

Audi, Robert (ed.) (1995), *The Cambridge Dictionary of Philosophy*, Cambridge and New York: Cambridge University Press.

Babich, Babette E. (2003), 'From Fleck's *Denkstil* to Kuhn's Paradigm: Conceptual Schemes and Incommensurability', *International Studies in the Philosophy of Science* 17: 1, pp. 75–92.

Baker, Steve (2002), 'What Does Becoming-Animal Look Like?', in Nigel Rothfels (ed.), *Representing Animals*, Bloomington: Indiana University Press, pp. 67–98.

Baker, Steve (2003), 'Sloughing the Human', in C. Wolfe (ed.), *Zoontologies: The Question of the Animal*, Minneapolis: University of Minnesota Press, pp. 147–64.

Barkow, Jerome H. (1989), *Darwin, Sex, and Status: Biological Approaches to Mind and Culture*, Toronto: University of Toronto Press.

Barkow, Jerome H. et al. (1992), 'Introduction: Evolutionary Psychology and Conceptual Integration', in Jerome H. Barkow et al. (eds), *The Adapted Mind: Evolutionary Psychology and the Generation of Culture*, Oxford: Oxford University Press, pp. 3–15.

Barkow, Jerome H., Leda Cosmides and John Tooby (eds) (1992), *The Adapted Mind: Evolutionary Psychology and the Generation of Culture*, Oxford: Oxford University Press.

Barnes, Barry (1982), *T. S. Kuhn and Social Science*, New York: Columbia University Press.

Barthes, Roland (1968), *Writing Degree Zero* and *Elements of Semiology*, both trans. Annette Lavers and Colin Smith, New York: Hill and Wang.

Bateson, Patrick (1987), 'Biological Approaches to the Study of Behavioral Development', *International Journal of Behavioral Development* 10: 1, pp. 1–22.

Bateson, Patrick (ed.) (1991), *The Development and Integration of Behaviour: Essays in Honour of Robert Hinde*, Cambridge: Cambridge University Press.

Bechtel, William (1988), *Philosophy of Mind: An Overview for Cognitive Science*, Hillsdale: Erlbaum.

Bechtel, William (1999), 'Unity of Science', in Robert A. Wilson and Frank C. Keil (eds), *The MIT Encyclopedia of the Cognitive Sciences*, Cambridge: MIT Press, pp. 856–7.

Becker, Carl Lotus (1935), *Everyman His Own Historian: Essays on History and Politics*, New York: F. S. Crofts.

Bennett, M. R., and P. M. S. Hacker (2003), *Philosophical Foundations of Neuroscience*, Malden: Blackwell.

Berger, Peter L., and Thomas Luckmann (1966), *The Social Construction of Reality: A Treatise in the Sociology of Knowledge*, Garden City: Doubleday.

Berry, Wendell (1977), *The Unsettling of America: Culture and Agriculture*, San Francisco: Sierra Club Books.

Biagioli, Mario (ed.) (1999), *The Science Studies Reader*, New York: Routledge.

Blackmore, Susan (1999), *The Meme Machine*, Oxford: Oxford University Press.

Blake, Andrew and Alan Yuille (eds) (1992), *Active Vision*, Cambridge: MIT Press.

Bloor, David [1976] (1991), *Knowledge and Social Imagery*, 2nd edn, Chicago: University of Chicago Press.

Bloor, David (1983), *Wittgenstein: A Social Theory of Knowledge*, New York: Columbia University Press.

Boghossian, Paul (1998), 'What the Sokal Hoax Ought to Teach Us', in N. Koertge (ed.), *A House Built on Sand: Exposing Postmodernist Myths about Science*, New York: Oxford University Press, pp. 23–31.

Boghossian, Paul (2002), 'Constructivist and Relativist Conceptions of Knowledge in Contemporary (Anti-)Epistemology: A Reply to Barbara Herrnstein Smith', in *South Atlantic Quarterly* 101: 1, pp. 213–28.

Borges, Jorge Luis (1966), *Other Inquisitions, 1937–1952*, trans. Ruth L. C. Simms, New York: Washington Square Press.

Bourdieu, Pierre (1984), *Distinction: A Social Critique of the Judgment of Taste*, trans. Richard Nice, Cambridge: Harvard University Press.

Bowker, Geoffrey and Susan Leigh Star (1999), *Sorting Things Out: Classification and its Consequences*, Cambridge: MIT Press.

Brint, Steven (ed.) (2002), *The Future of the City of Intellect: The Changing American University*, Stanford: Stanford University Press.

Brooks, Rodney A. (1991), 'Intelligence without Representation', *Artificial Intelligence* 47: pp. 139–59.

Brorson, Stig (2000), *On the Socio-Cultural Preconditions of Medical Cognition: Studies in Ludwik Fleck's Medical Epistemology*, Ph.D. dissertation, Department of Medical Philosophy and Clinical Theory, The Faculty of Health Sciences, University of Copenhagen.

Brown, Wendy (1995), *States of Injury: Power and Freedom in Late Modernity*, Princeton: Princeton University Press.

Burrow, J. W. (2000), *The Crisis of Reason: European Thought, 1848–1914*, New Haven: Yale University Press.

Buss, David M. (1994), *The Evolution of Desire: Strategies of Human Mating*, New York: Basic Books.

Buss, David M. (2000), *The Dangerous Passion: Why Jealousy is as Necessary as Love and Sex*, New York: Free Press.

Calasso, Roberto (1993), *The Marriage of Cadmus and Harmony*, trans. Tim Parks, New York: Knopf.

Caneva, Kenneth L (1998), 'Objectivity, Relativism and the Individual: A Role for a Post-Kuhnian History of Science,' *Studies in the History and Philosophy of Science*, 29: 3, pp. 330–1.

Carroll, Joseph (2004), *Literary Darwinism: Evolution, Human Nature, and Literature*, New York: Routledge.

Changeux, Jean-Pierre [1983] (1997), *Neuronal Man: The Biology of Mind*, trans. Laurence Garey, Princeton: Princeton University Press.

Changeux, Jean-Pierre, and Paul Ricoeur (2000), *What Makes Us Think?: A Neuroscientist and a Philosopher Argue about Ethics, Human Nature, and the Brain*, trans. M. B. DeBevoise, Princeton: Princeton University Press.

Cheney, Dorothy L. and Richard M. Seyfarth (1990), *How Monkeys See the World*, Chicago: University of Chicago Press.

Christensen, W. D. and C. A. Hooker (2000), 'An Interactionist-Constructivist Approach to Intelligence: Self-Directed Anticipative Learning', *Philosophical Psychology* 13: 1, pp. 5–45.

Cixous, Hélène (1998), 'Love of the Wolf', in Hélène Cixous, *Stigmata: Escaping Texts*, London: Routledge, pp. 84–99.

Clark, Andy (1993), *Associative Engines: Connectionism, Concepts, and Representational Change*, Cambridge: MIT Press.

Clark, Andy (1997), *Being There: Putting Brain, Body, and World Together Again*, Cambridge: MIT Press.

Clark, Andy (1998), 'Time and Mind', *The Journal of Philosophy*, 95: 7, pp. 354–76.

Code, Lorraine (1993), 'Taking Subjectivity into Account', in Linda Alcoff and Elizabeth Potter (eds), *Feminist Epistemologies*, New York: Routledge, pp.15–48.

Cohen, Robert S. and Thomas Schnelle (eds) (1986), *Cognition and Fact: Materials on Ludwik Fleck*, Dordrecht: D. Reidel.

Collini, Stefan (1993), 'Introduction', in C. P. Snow, *The Two Cultures*, Cambridge: Cambridge University Press.

Collins, Harry and Trevor Pinch (1993), *The Golem: What Everyone should Know about Science*, Cambridge: Cambridge University Press.

Collins, Randall (1998), *The Sociology of Philosophies: A Theory of Global Intellectual Change*, Cambridge: Harvard University Press.

Connolly, William E. (1991), *Identity/Difference: Democratic Negotiations of Political Paradox*, Ithaca: Cornell University Press.

Connolly, William E. (1995), *The Ethos of Pluralization*, Minneapolis: University of Minnesota Press.

Connolly, William (2005), *Neuropolitics: Thinking, Culture, Speed*, Minneapolis: University of Minnesota Press.

Cox, Graham and Tony Ashford (1998), 'Riddle Me This: The Craft and Concept of Animal Mind', *Science, Technology and Human Values*, 23: 4, pp. 425–38.

Crawford, Charles and Dennis L. Krebs (eds) (1998), *Handbook of Evolutionary Psychology: Ideas, Issues, and Applications*, Mahwah: Erlbaum.
Culler, Jonathan (1975), *Structuralist Poetics*, Ithaca: Cornell University Press.
Cussins, Charis (1996), 'Ontological Choreography: Agency through Objectification in Infertility Clinics', *Social Studies of Science*, 26: 3, pp. 575–610.
Damasio, Antonio R. (1994), *Descartes' Error: Emotion, Reason, and the Human Brain*, New York: Avon.
Davidson, Donald [1974] (1985), 'On the Very Idea of a Conceptual Scheme', in Davidson, *Inquiries into Truth and Interpretation*, Oxford: Clarendon Press, pp. 183–98.
Davidson, Donald [1988] (2002), 'Epistemology and Truth', in Davidson, *Subjective, Intersubjective, Objective*, Oxford: Oxford University Press, pp. 177–92.
Dawkins, Richard (1976), *The Selfish Gene*, New York: Oxford University Press.
Deacon, Terrence W. (1997), *The Symbolic Species: The Co-Evolution of Language and the Brain*, New York: Norton.
Deleuze, Gilles and Félix Guattari (1987), 'Becoming-Animal', in Deleuze and Guattari, *A Thousand Plateaus: Capitalism and Schizophrenia*, trans. Brian Massumi, Minneapolis: University of Minnesota Press, pp. 232–309.
Derrida, Jacques (1976), *Of Grammatology*, trans. Gayatri Chakravorty Spivak, Baltimore: Johns Hopkins University Press.
Derrida, Jacques (1990), 'Force of Law: The "Mystical Foundation of Authority"', trans. Mary Quaintance, *Cardozo Law Review*, 11: 19, pp. 919–1,045.
Derrida, Jacques (1991), '"Eating Well" or the Calculation of the Subject', trans. Peter Connor and Avital Ronell, in Eduardo Cadava, Peter Connor and Jean-Luc Nancy (eds), *Who Comes after the Subject?*, New York: Routledge, pp. 96–119.
Derrida, Jacques (2001), *On Cosmopolitanism and Forgiveness*, trans. Mark Dooley and Michael Hughes, New York: Routledge.
Derrida, Jacques (2003), 'And Say the Animal Responded?', trans. David Wills, in C. Wolfe (ed.), *Zoontologies: The Question of the Animal*, Minneapolis: University of Minnesota Press, pp. 121–46.
Diamond, Cora (1994), 'Truth: Defenders, Debunkers, Despisers', in Leona Toker (ed.), *Commitment in Reflection: Essays in Literature and Moral Philosophy*, Hamden: Garland, pp. 195–222.
Dilthey, Wilhelm [1910] (2002), *The Formation of the Historical World in the Human Sciences*, ed. Rudolf A. Makkreel and Frithjof Rodi, Princeton: Princeton University Press.
Dupré, John (1993), *The Disorder of Things: Metaphysical Foundations of the Disunity of Science*, Cambridge: Harvard University Press.
Dupré, John (2001), *Human Nature and the Limits of Science*, Oxford: Clarendon Press.
Eagleton, Terry (1990), *The Ideology of the Aesthetic*, Oxford: Blackwell.
Easterlin, Nancy (ed.) (2001), *Symposium: Evolution and Literature*, spec. issue of *Philosophy and Literature* 25, pp. 197–346.
Easton, David and Corinne S. Schelling (eds) (1991), *Divided Knowledge: Across Disciplines, Across Cultures*, Newbury Park: Sage Publications.

Edelman, Gerald (1989), *The Remembered Present: A Biological Theory of Consciousness*, New York: Basic Books.

Edelman, Gerald (1992), *Bright Air, Brilliant Fire: On the Matter of Mind*, New York: Basic Books.

Epstein, Steven (1996), *Impure Science: AIDS, Activism, and the Politics of Knowledge*, Berkeley: University of California Press.

Ereshefsky, Marc (1998), 'Species Pluralism and Anti-Realism', *Philosophy of Science* 65: 1, pp. 103–20.

Evans, Richard J. (1999), *In Defense of History*, New York: Norton.

Fernández-Armesto, Felipe (1997), *Truth: A History and Guide for the Perplexed*, New York: St. Martin's Press.

Ferry, Luc (1995), *The New Ecological Order*, trans. Carol Volk, Chicago: University of Chicago Press.

Festinger, Leon [1957] (1962), *A Theory of Cognitive Dissonance*, Stanford: Stanford University Press.

Feyerabend, Paul (1975), *Against Method: Outline of an Anarchistic Theory of Knowledge*, Atlantic Highlands: Humanities Press.

Fleck, Ludwik [1935] (1979), *Genesis and Development of a Scientific Fact*, ed. Thaddeus J. Trenn and Robert K. Merton, trans. Fred Bradley and Thaddeus J. Trenn, Chicago: University of Chicago Press.

Fleck, Ludwik [1935] (1980), *Entstehung und Entwicklung einer wissenschaftlichen Tatsache*, Frankfurt am Main: Suhrkamp Verlag.

Fodor, Jerry A. (1981), *Representations: Philosophical Essays on the Foundations of Cognitive Science*, Cambridge: MIT Press.

Fodor, Jerry A. (1998), *Concepts: Where Cognitive Science Went Wrong*, Oxford: Clarendon Press.

Foucault, Michel (1971), *The Order of Things: An Archaeology of the Human Sciences*, trans. A. M. Sheridan Smith, New York: Pantheon.

Foucault, Michel (1972), *The Archaeology of Knowledge*, trans. A. M. Sheridan Smith, New York: Pantheon.

Foucault, Michel (1979), *Discipline and Punish: The Birth of the Prison*, trans. Alan Sheridan, New York: Vintage.

Frankenheimer, John (dir.) (1996), *The Island of Dr Moreau*, New Line Cinema.

Friedman, Michael (2000), *A Parting of the Ways: Carnap, Cassirer, and Heidegger*, Chicago: Open Court.

Fuller, Steve (2000), *Thomas Kuhn: A Philosophical History for Our Times*, Chicago: University of Chicago Press.

Galison, Peter and David J. Stump (eds) (1996), *The Disunity of Science: Boundaries, Contexts, and Power*, Stanford: Stanford University Press.

Geertz, Clifford (1984), 'Anti-Anti-Relativism', *American Anthropologist* 86: 2, pp. 263–78.

Gibson, James J. (1966), *The Senses Considered as Perceptual Systems*, Boston: Houghton Mifflin.

Gibson, James J. (1979), *The Ecological Approach to Visual Perception*, Boston: Houghton Mifflin.

Gibson, Kathleen R. and Tim Ingold (1993), *Tools, Language and Cognition in Human Evolution*, Cambridge: Cambridge University Press.

Giere, Ronald (1999), *Science without Laws*, Chicago: University of Chicago Press.

Gieryn, Thomas F. (2004), 'Eloge: Robert K. Merton, 1910–2003', *Isis* 95: 1, pp. 91–4.

Godfrey-Smith, Peter (2003), *Theory and Reality: An Introduction to the Philosophy of Science*, Chicago: University of Chicago Press.

Golinski, Jan (1998), *Making Natural Knowledge: Constructivism and the History of Science*, Cambridge: Cambridge University Press.

Goodall, Jane (1986), *The Chimpanzees of Gombe: Patterns of Behavior*, Cambridge: Harvard University Press.

Goodman, Nelson (1978), *Ways of Worldmaking*, Indianapolis: Hackett.

Gould, S. J. and R. C. Lewontin (1979), 'The Spandrels of San Marco and the Panglossian Paradigm: A Critique of the Adaptationist Programme', *Proceedings of the Royal Society of London: Series B, Biological Sciences*, 205:1161, pp. 581–98.

Greenwood, John D. (2004), *The Disappearance of the Social in American Social Psychology*, Cambridge: Cambridge University Press.

Griffin, Donald R. (1984), *Animal Thinking*, Cambridge: Harvard University Press.

Griffiths, Paul E. (1997), *What Emotions Really Are: The Problem of Psychological Categories*, Chicago: University of Chicago Press.

Griffiths, Paul E. (2001), 'From Adaptive Heuristic to Phylogenetic Perspective: Some Lessons from the Evolutionary Psychology of Emotion', in Harmon R. Holcomb, III (ed.), *Conceptual Challenges in Evolutionary Psychology: Innovative Research Strategies*, Dordrecht: Kluwer, pp. 309–26.

Gross, Paul R. and Norman Levitt (1994), *Higher Superstition: The Academic Left and its Quarrels with Science*, Baltimore: Johns Hopkins University Press.

Gross, Paul R., Norman Levitt and Martin W. Lewis (eds) (1996), *The Flight from Science and Reason*, New York: New York Academy of Sciences.

Guillory, John (2002), 'The Sokal Affair and the History of Criticism', *Critical Inquiry* 28: 2, pp. 470–508.

Haack, Susan (1996), 'Concern for Truth: What It Means, Why It Matters', in Paul R. Gross, Norman Levitt and Martin W. Lewis (eds), *The Flight from Science and Reason*, New York: New York Academy of Sciences, pp. 57–63.

Haack, Susan (2003), *Defending Science – Within Reason: Between Scientism and Cynicism*, Amherst: Prometheus.

Hacking, Ian (1992), 'The Self-Vindication of the Laboratory Sciences', in Andrew Pickering (ed.), *Science as Practice and Culture*, Chicago: University of Chicago Press.

Hacking, Ian (1999), *The Social Construction of* What?, Cambridge: Harvard University Press.

Hansen, Mark B. N. (2004), *New Philosophy for New Media*, Cambridge: MIT Press.

Haraway, Donna (1991), *Simians, Cyborgs and Women: The Reinvention of Nature*, New York: Routledge.

Haraway, Donna (1997), *Modest_Witness@Second-Millennium.Female Man©_Meets_OncoMouse™: Feminism and Technoscience*, New York: Routledge.

Harari, Josué V. (ed.) (1979), *Textual Strategies: Perspectives in Post-Structuralist Criticism*, Ithaca: Cornell University Press.

Hardcastle, Gary L. and Alan W. Richardson (eds) (2003), *Logical Empiricism in North America*, Minneapolis: University of Minnesota Press.

Harding, Sandra (ed.) (1993a), *The 'Racial' Economy of Science: Toward a Democratic Future*, Bloomington: Indiana University Press.

Harding, Sandra (1993b), 'Rethinking Standpoint Epistemology: What is "Strong Objectivity"?', in Linda Alcoff and Elizabeth Potter (eds), *Feminist Epistemologies*, New York: Routledge.

Harding, Sandra (1998), *Is Science Multicultural? Postcolonialisms, Feminisms, and Epistemologies*, Indianapolis: Indiana University Press.

Harré, Rom and Michael Krausz (1996), *Varieties of Relativism*, Oxford: Blackwell.

Hayek, Friedrich A. von (1936), 'Economics and Knowledge', Presidential address delivered before the London Economics Club in 1936, *Economica* IV (new series, 1937), pp. 33–54.

Heidegger, Martin [1952] (1977), 'The Age of the World Picture', in *The Question Concerning Technology and Other Essays*, trans. William Lovitt, New York: Harper and Row.

Hendriks-Jansen, Horst (1996), *Catching Ourselves in the Act: Situated Activity, Interactive Emergence, Evolution, and Human Thought*, Cambridge: MIT Press.

Herbert, Christopher (2001), *Victorian Relativity: Radical Thought and Scientific Discovery*, Chicago: University of Chicago Press.

Himmelfarb, Gertrude (1994), *On Looking into the Abyss: Untimely Thoughts on Culture and Society*, New York: Knopf.

Høeg, Peter (1996), *The Woman and the Ape*, trans. Barbara Haveland, New York: Farrar.

Hofstadter, Douglas (1998), 'Popular Culture and the Threat to Rational Inquiry', *Science* 281: 5,376, pp. 512–13.

Holcomb, Harmon R., III (ed.) (2001), *Conceptual Challenges in Evolutionary Psychology: Innovative Research Strategies*, Dordrecht: Kluwer.

Hollis, Martin and Steven Lukes (eds) (1982), *Rationality and Relativism*, Cambridge: MIT Press.

Hoyningen-Huene, Paul (1993), *Reconstructing Scientific Revolutions: Thomas S. Kuhn's Philosophy of Science*, trans. Alexander T. Levine, Chicago: University of Chicago Press.

Huen, Chi W. (2001), *In the Wake of the Science Wars: An Experiment with the Anthropology of the Academy*, Ph.D. dissertation, Department of Archeology and Anthropology, Cambridge University.

Hutchins, Edwin (1995), *Cognition in the Wild*, Cambridge: MIT Press.

Ingold, Tim (1993), 'Technology, Language, Intelligence: A Reconsideration of Basic Concepts', in K. R. Gibson and T. Ingold (eds), *Tools, Language, and Cognition in Human Evolution*, Cambridge: Cambridge University Press, pp. 449–72.

Ingold, Tim (2000), 'Evolving Skills', in Hilary Rose and Steven Rose (eds), *Alas, Poor Darwin: Arguments against Evolutionary Psychology*, New York: Harmony, pp. 273–98.

Jameson, Fredric (1984), 'Postmodernism, or, The Cultural Logic of Late Capitalism', *New Left Review* 146, pp. 59–92.

Jamieson, Dale (2002), *Morality's Progress: Essays on Humans, Other Animals, and the Rest of Nature*, Oxford: Clarendon Press.

Jay, Martin (2004), 'Modernism and the Specter of Psychologism', in Mark S. Micale (ed.), *The Mind of Modernism: Medicine, Psychology, and the Cultural Arts in Europe and America, 1880–1940*, Stanford: Stanford University Press, pp. 352–66.

Jones, James H. (1981), *Bad Blood: The Tuskegee Syphilis Experiment*, New York: Free Press.

Karmiloff-Smith, Annette (2000), 'Why Babies' Brains Are Not Swiss Army Knives', in Hilary Rose and Steven Rose (eds), *Alas, Poor Darwin: Arguments against Evolutionary Psychology*, London: Jonathan Cape, pp. 173–88.

Karp, David L. (2003), *Shakespeare, Einstein, and the Bottom Line: The Marketing of Higher Education*, Cambridge: Harvard University Press.

Keynes, John Maynard (1936), *The General Theory of Employment, Interest, and Money*, New York: Harcourt, Brace.

Kitcher, Philip (1993), *The Advancement of Science: Science without Legend, Objectivity without Illusions*, New York: Oxford University Press.

Knorr-Cetina, Karin (1981), *The Manufacture of Knowledge: An Essay on the Constructivist and Contextual Nature of Science*, Oxford: Pergamon Press.

Knorr-Cetina, Karin (1999), *Epistemic Cultures: How the Sciences Make Knowledge*, Cambridge: Harvard University Press.

Koertge, Noretta (1998a), 'Scrutinizing Science Studies', in Noretta Koertge (ed.), *A House Built on Sand: Exposing Postmodernist Myths about Science*, New York: Oxford University Press, pp. 3–6.

Koertge, Noretta (ed.) (1998b), *A House Built on Sand: Exposing Postmodernist Myths about Science*, New York: Oxford University Press.

Kuhn, Thomas [1962] (1996), *The Structure of Scientific Revolutions*, 3rd edn, Chicago: University of Chicago Press.

Kuhn, Thomas (1979), 'Foreword', in Ludwik Fleck, *Genesis and Development of a Scientific Fact*, ed. Thaddeus J. Trenn and Robert K. Merton, trans. Fred Bradley and Thaddeus J. Trenn, Chicago: University of Chicago Press, pp. vii–xii.

Kuhn, Thomas (1989), 'Response to Commentators', in S. Allén (ed.), *Possible Worlds in the Humanities, Arts, and Sciences*, Berlin: De Gruyter, pp. 49–51.

Kusch, Martin (1995), *Psychologism: A Case Study in the Sociology of Philosophical Knowledge*, London: Routledge.

Labinger, Jay and Harry Collins (eds) (2001), *The One Culture?: A Conversation about Science*, Chicago: University of Chicago Press.

Laclau, Ernesto and Chantal Mouffe (1987), *Hegemony and Socialist Strategy: Towards a Radical Democratic Politics*, London: Verso.

Lakoff, George (1987), *Women, Fire, and Dangerous Things: What Categories Reveal about the Mind*, Chicago: University of Chicago Press.

Lakoff, George and Mark Johnson (1980), *Metaphors We Live By*, Chicago: University of Chicago Press.

Laqueur, Thomas (1990), *Making Sex: Body and Gender from the Greeks to Freud*, Cambridge: Harvard University Press.

Latour, Bruno (1987), *Science in Action: How to Follow Scientists and Engineers through Society*, Cambridge: Harvard University Press.

Latour, Bruno (1988), *The Pasteurization of France*, trans. Alan Sheridan and John Law, Cambridge: Harvard University Press.

Latour, Bruno (1993), *We Have Never Been Modern*, trans. Catherine Porter, Cambridge: Harvard University Press.

Latour, Bruno (1999), *Pandora's Hope: Essays on the Reality of Science Studies*, Cambridge: Harvard University Press.

Latour, Bruno (2004), *Politics of Nature: How to Bring the Sciences into Democracy*, Cambridge: Harvard University Press.

Latour, Bruno and Steve Woolgar (1986), *Laboratory Life: The [Social] Construction of Scientific Facts*, Princeton: Princeton University Press.

Laudan, Larry (1996), *Beyond Positivism and Relativism: Theory, Method, and Evidence*, Boulder: Westview Press.

Leach, Edmund (1982), *Social Anthropology*, New York: Oxford University Press.

Leahey, Thomas H. (1980), *A History of Psychology: Main Currents in Psychological Thought*, Englewood Cliffs: Prentice-Hall.

Lepenies, Wolf (1988), *Between Literature and Science: The Rise of Sociology*, trans. R. J. Hollingdale, Cambridge: Cambridge University Press.

Lieberman, Philip (1998), *Eve Spoke: Human Language and Human Evolution*, New York: Norton.

Lipstadt, Deborah (1993), *Denying the Holocaust: The Growing Assault on Truth and Memory*, New York: The Free Press.

Llewelyn, John (1991), *The Middle Voice of Ecological Conscience: A Chiasmic Reading of Responsibility in the Neighborhood of Levinas, Heidegger, and Others*, New York: St. Martin's Press.

Longino, Helen E. (2002), *The Fate of Knowledge*, Princeton: Princeton University Press.

Luckhurst, Nicola (2000), *Science and Structure in Proust's* A la recherche du temps perdu, Oxford: Clarendon Press.

Lyotard, Jean-François (1984), *The Postmodern Condition: A Report on Knowledge*, trans. Geoff Bennington and Brian Massumi, Minneapolis: University of Minnesota Press.

McCauley, Robert N. (ed.) (1996), *The Churchlands and Their Critics*, Cambridge: Blackwell.

McCumber, John (2001), *Time in the Ditch: American Philosophy and the McCarthy Era*, Evanston: Northwestern University Press.

Machlup, Fritz (1982), *Knowledge: Its Creation, Distribution, and Economic Significance, vol. II, The Branches of Learning*, Princeton: Princeton University Press.

Magnus, P. D. and Craig Callendar (2003), 'Realist Ennui and the Base Rate Fallacy', *Philosophy of Science* 71: 3, pp. 320–38.

Mallon, Ron and Stephen P. Stich (2000), 'The Odd Couple: The Compatibility of Social Construction and Evolutionary Psychology', *Philosophy of Science* 67, pp. 133–54.

Mandelbaum, Maurice [1938] (1967), *The Problem of Historical Knowledge: An Answer to Relativism*, New York: Harper and Row.

Maturana, Humberto R. (1991), 'Reality', in Norbert Leser, Josef Seifert and Klaus Plitzner (eds), *Die Gedank Sir Karl Poppers: Kritischer Rationalismus im Dialog*, no translator given, Heidelberg: Carl Winter Universitätsverlag, pp. 282–357.

Maturana, Humberto R. and Francisco J. Varela (1980), *Autopoiesis and Cognition: The Realization of the Living*, Boston: D. Reidel.
Maturana, Humberto R. and Francisco J. Varela (1992), *The Tree of Knowledge: The Biological Roots of Human Understanding*, Boston: Shambala.
Mayr, Ernst (1988), *Toward a New Philosophy of Biology*, Cambridge: Harvard University Press.
Menand, Louis (2001), *The Metaphysical Club: A Story of Ideas in America*, New York: Farrar, Straus and Giroux.
Mohanty, Satya (1997), *Literary Theory and the Claims of History: Postmodernism, Objectivity, Multicultural Politics*, Ithaca: Cornell University Press.
Nagel, Thomas (1998), 'The Sleep of Reason', *The New Republic* (12 October), pp. 32–8.
Nelson, Richard (1987), 'Carl Becker Revisited: Irony and Progress in History', *Journal of the History of Ideas*, 48: 2, pp. 307–23.
Newfield, Christopher (2003), *Ivy and Industry: Business and the Making of the American University, 1880–1980*, Durham: Duke University Press.
Nickles, Thomas (2000), 'Discovery', in W. H. Newton-Smith (ed.), *A Companion to the Philosophy of Science*, Oxford: Blackwell, pp. 85–96.
Nickles, Thomas (2003), 'Normal Science: From Logic to Case-Based and Model-Based Reasoning', in Thomas Nickles (ed.), *Thomas Kuhn*, Cambridge: Cambridge University Press, pp.142–77.
Nisbett, R. E. and L. Ross (1980), *Human Inference: Strategies and Shortcomings of Social Judgment*, Englewood Cliffs, NJ: Prentice Hall.
Norris, Christopher (1990), *What's Wrong with Postmodernism: Critical Theory and the Ends of Philosophy*, Baltimore: Johns Hopkins University Press.
Norris, Christopher (1992), *Uncritical Theory: Postmodernism, Intellectuals and the Gulf War*, Amherst: University of Massachusetts Press.
Notturno, Mark A. (ed.) (1989), *Perspectives on Psychologism*, Leiden: E. J. Brill.
Novick, Peter (1988), *That Noble Dream: The 'Objectivity Question' and the American Historical Profession*, Cambridge: Cambridge University Press.
Nozick, Robert (2001), *Philosophical Explanations*, Cambridge: Harvard University Press.
Odling-Smee, F. John, Kevin N. Laland and Marcus W. Feldman (1996), 'Niche Construction', *The American Naturalist* 147: 4, pp. 641–8.
Oleson, Alexandra and John Voss (eds) (1979), *The Organization of Knowledge in Modern America, 1860–1920*, Baltimore: Johns Hopkins University Press.
Oyama, Susan [1985] (2000), *The Ontogeny of Information: Developmental Systems and Evolution*, 2nd edn, Durham: Duke University Press.
Oyama, Susan (2000), *Evolution's Eye: A Systems View of the Biology-Culture Divide*, Durham: Duke University Press.
Oyama, Susan, Paul E. Griffiths and Russell Gray (eds) (2001), *Cycles of Contingency: Developmental Systems and Evolution*, Cambridge: MIT Press.
Petitot, Jean, Francisco J. Varela, Bernard Pachout, and Jean-Michel Roy (eds) (1999), *Naturalizing Phenomenology: Issues in Contemporary Phenomenology and Cognitive Science*, Stanford: Stanford University Press.
Piaget, Jean (1955), *The Construction of Reality in the Child*, trans. Margaret Cook, London: Routledge and Kegan Paul.

Piaget, Jean (1971), *Biology and Knowledge: An Essay on the Relations between Organic Regulations and Cognitive Processes*, trans. Beatrix Walsh, Chicago: University of Chicago Press.

Pickering, Andrew (1984), *Constructing Quarks: A Sociological History of Particle Physics*, Chicago: University of Chicago Press.

Pickering, Andrew (1995), *The Mangle of Practice: Time, Agency, and Science*, Chicago: University of Chicago Press.

Pinker, Steven (1994), *The Language Instinct*, New York: W. Morrow and Co.

Pinker, Steven (1997), *How the Mind Works*, New York: Norton.

Pinker, Steven (2002), *The Blank Slate: The Modern Denial of Human Nature*, New York and London: Viking Penguin.

Pinker, Steven and Paul Bloom (1992), 'Natural Language and Natural Selection', in J. H. Barkow et al. (eds), *The Adapted Mind: Evolutionary Psychology and the Generation of Culture*, Oxford: Oxford University Press, pp. 451–94.

Pippin, Robert (1999), *Modernism as a Philosophical Problem: On the Dissatisfactions of European High Culture*, 2nd edn, Malden: Blackwell.

Plotkin, Henry (1998), *Evolution in Mind: An Introduction to Evolutionary Psychology*, Cambridge: Harvard University Press.

Polyani, Michael (1958), *Personal Knowledge: Towards a Post-Critical Philosophy*, Chicago: University of Chicago Press.

Popper, Karl [1935] (1965), *The Logic of Scientific Discovery*, rev. edn, no trans. given, New York: Harper and Row.

Port, Robert F. and Timothy van Gelder (eds) (1995), *Mind as Motion: Explorations in the Dynamics of Cognition*, Cambridge: MIT Press.

Proctor, Robert N. (1991), *Value-Free Science?: Purity and Power in Modern Knowledge*, Cambridge: Harvard University Press.

Proust, Marcel [1927] (1981), *Remembrance of Things Past*, trans. C. K. Scott Moncrieff, Terence Kilmartin and Andreas Mayor, New York: Vintage.

Purves, Dale and R. Beau Lotto (2003), *Why We See What We Do: An Empirical Theory of Vision*, Sunderland: Sinnauer Associates.

Quine, W. V. (1969), 'Epistemology Naturalized', in W. V. Quine, *Ontological Relativity, and Other Essays*, New York: Columbia University Press, pp. 69–90.

Reed, Edward S. (1997), *From Soul to Mind: The Emergence of Psychology, from Erasmus Darwin to William James*, New Haven: Yale University Press.

Regan, Tom (1983), *The Case for Animal Rights*, Berkeley: University of California Press.

Rescher, Nicholas, 'Idealism', in Robert Audi (ed.) (1995), *The Cambridge Dictionary of Philosophy*, pp. 355–7.

Richardson, Alan and Francis F. Steen (eds) (2002), *Literature and the Cognitive Revolution*, spec. issue of *Poetics Today*, 23: 1, pp. 1–179.

Richardson, Robert C. (2001), 'Evolution without History: Critical Reflections on Evolutionary Psychology', in Holcomb, Harmon R., III (ed.), *Conceptual Challenges in Evolutionary Psychology: Innovative Research Strategies*, Dordrecht: Kluwer, pp. 327–73.

Ridderinkhof, K. R., M. Ullsperger and E. A. Crone (2004), 'The Role of the Medial Frontal Cortex in Cognitive Control', *Science* 306: 5,695, spec. issue on 'Cognition and Behavior', pp. 443–7.

Riordan, Michael (1997), 'Trading Information', rev. of Peter Galison, *Image and Logic: A Material Culture of Microphysics* (Chicago: University of Chicago Press, 1997), *New York Times Book Review*, 14 September, p. 38.

Ristau, Carolyn A. (1999), 'Cognitive Ethology', in Robert A. Wilson and Frank C. Keil (eds), *The MIT Encyclopedia of the Cognitive Sciences*, Cambridge: MIT Press, pp. 132–4.

Ritvo, Harriet (1997), *The Platypus and the Mermaid, and Other Figments of the Classifying Imagination*, Cambridge: Harvard University Press.

Romanes, George John [1884] (1969), *Mental Evolution in Animals*, New York: AMS Press.

Rooney, Phyllis (1998), 'Putting Naturalized Epistemology to Work', in Linda Martin Alcoff (ed.), *Epistemology: The Big Questions*, Oxford: Blackwell, pp. 285–305.

Rorty, Richard (1979), *Philosophy and the Mirror of Nature*, Princeton: Princeton University Press.

Rorty, Richard (1991), *Objectivity, Relativism, and Truth*, Cambridge: Cambridge University Press.

Rose, Hilary and Steven Rose (eds) (2000), *Alas, Poor Darwin: Arguments against Evolutionary Psychology*, New York: Harmony.

Rothstein, Edward (2001), 'Attacks on U.S. Challenge Postmodern True Believers', *New York Times*, 22 September, p. A17.

Rouse, Joseph (2002), 'Vampires: Social Constructivism, Realism, and Other Philosophical Undead', *History and Theory* 41: 1, pp. 60–78.

Savage-Rumbaugh, Sue, Stuart G. Shanker and Talbot Taylor (1998), *Apes, Language, and the Human Mind*, New York: Oxford University Press.

Scheffler, Israel (1967), *Science and Subjectivity*, Indianapolis: Bobbs-Merrill.

Schnelle, Thomas (1986), 'Microbiology and Philosophy of Science, Lwów and the German Holocaust: Stations of a Life – Ludwik Fleck 1896–1961', in Robert S. Cohen and Thomas Schnelle (eds), *Cognition and Fact: Materials on Ludwik Fleck*, Dordrecht: D. Reidel, pp. 19–29.

Serres, Michel, with Bruno Latour [1990] (1995), *Conversations on Science, Culture, and Time*, trans. Roxanne Lapidus, Ann Arbor: University of Michigan Press.

Shapin, Steven and Simon Schaffer (1985), *Leviathan and the Air Pump: Hobbes, Boyle, and the Experimental Life*, Princeton: Princeton University Press.

Sheets-Johnstone, Maxine (1990), *The Roots of Thinking*, Philadelphia: Temple University Press.

Sheets-Johnstone, Maxine (1998), 'Consciousness: A Natural History', *Journal of Consciousness Studies*, 5: 3, pp. 260–94.

Shiebinger, Londa (1993), *Nature's Body: Gender in the Making of Modern Science*, Boston: Beacon Press.

Singer, Peter (1990), *Animal Liberation*, New York: Avon.

Smith, Barbara Herrnstein (1988), *Contingencies of Value: Alternative Perspectives for Critical Theory*, Cambridge: Harvard University Press.

Smith, Barbara Herrnstein (1997), *Belief and Resistance: Dynamics of Contemporary Intellectual Controversy*, Cambridge: Harvard University Press.

Smith, Barbara Herrnstein (2002), 'Reply to an Analytic Philosopher', *South Atlantic Quarterly* 101: 1, pp. 229–42.

Smith, Barbara Herrnstein and Arkady Plotnitsky (eds) (1997), *Mathematics, Science, and Postclassical Theory*, Durham: Duke University Press.

Smith, Brian Cantwell (1996), *On the Origin of Objects*, Cambridge: MIT Press.

Snow, C. P. [1959] (1998), *The Two Cultures*, Cambridge: Cambridge University Press.

Sober, Elliott (1998), 'Morgan's Canon', in Colin Allen and Denise Cummins (eds), *The Evolution of Mind*, New York: Oxford University Press, pp. 224–43.

Sokal, Alan (2000), *The Sokal Hoax: The Sham that Shook the Academy*, Lincoln: University of Nebraska Press.

Sokal, Alan and Jean Bricmont (1998), *Fashionable Nonsense: Postmodern Intellectuals' Abuse of Science*, New York: Picador.

Spivak, Gayatri Chakravorty (1988), *In Other Worlds: Essays in Cultural Politics*, New York: Routledge.

Spivak, Gayatri Chakravorty (1993), *Outside in the Teaching Machine*, New York: Routledge.

Stengers, Isabelle (2000), *The Invention of Modern Science*, Minneapolis: University of Minneapolis Press.

Sterelny, Kim and Paul E. Griffiths (1999), *Sex and Death: An Introduction to Philosophy of Biology*, Chicago: University of Chicago Press.

Stolberg, Michael (2003), 'A Woman Down to Her Bones: The Anatomy of Sexual Difference in the Sixteenth and Early Seventeenth Centuries', *Isis* 94: 2, pp. 274–99.

Stolberg, Sheryl Gay (2002), 'Bush's Advisers on Ethics Discuss Human Cloning', *New York Times*, 18 January, p. A18.

Thelen, Esther and Linda B. Smith (1994), *A Dynamic Systems Approach to the Development of Cognition and Action*, Cambridge: MIT Press.

Thornhill, Randy and Craig T. Palmer (2000), *A Natural History of Rape: Biological Bases of Sexual Coercion*, Cambridge: MIT Press.

Tomasello, Michael, Sue Savage-Rumbaugh and Ann Cale Kruger (1993), 'Imitative Learning of Actions on Objects by Children, Chimpanzees and Enculturated Chimpanzees', *Child Development*, 64: 6, pp. 1,688–705.

Tooby, John and Leda Cosmides (1992), 'The Psychological Foundations of Culture', in J. H. Barkow et al. (eds), *The Adapted Mind: Evolutionary Psychology and the Generation of Culture*, Oxford: Oxford University Press, pp. 19–136.

Trenn, T. J. and Robert Merton (1979), 'Biographical Sketch', in Ludwik Fleck, *Genesis and Development of a Scientific Fact*, ed. Thaddeus J. Trenn and Robert K. Merton, trans. Fred Bradley and Thaddeus J. Trenn, Chicago: University of Chicago Press, pp. 149–53.

van den Belt, Henk (1997), *Spirochaetes, Serology and Salvarsan: Ludwik Fleck and the Construction of Medical Knowledge about Syphilis*, Ph.D. dissertation, Catholic University of Nijmegen.

van Gelder, Timothy (1995a), 'What Might Cognition Be, If Not Computation?', *Journal of Philosophy*, 92: 7, pp. 345–81.

van Gelder, Timothy (1995b), 'It's About Time: An Overview of the Dynamical Approach to Cognition', in Robert F. Port and Timothy van Gelder (eds), *Mind as Motion: Explorations in the Dynamics of Cognition*, Cambridge: MIT Press, pp. 1–44.

Varela, Francisco J., Evan Thompson and Eleanor Rosch (1991), *The Embodied Mind: Cognitive Science and Human Experience*, Cambridge: MIT Press.

Vargish, Thomas and Delo E. Mook (1999), *Inside Modernism: Relativity Theory, Cubism, Narrative*, New Haven: Yale University Press.

von Glasersfeld, Ernst (1995), *Radical Constructivism: A Way of Knowing and Learning*, London: The Falmer Press.

de Waal, Frans (1996), *Good Natured: The Origins of Right and Wrong in Humans and Other Animals*, Cambridge: Harvard University Press.

Wheatcroft, Geoffrey (2001), 'Bearing False Witness', *New York Times Book Review*, 13 May, pp. 12–13.

White, Hayden (1973), *Metahistory: The Historical Imagination in Nineteenth-Century Europe*, Baltimore: Johns Hopkins University Press.

Williams, Bernard (1991), 'Prologue: Making Sense of Humanity', in James J. Sheehan and Morton Sosna (eds), *The Boundaries of Humanity: Humans, Animals, Machines*, Berkeley: University of California Press, pp. 13–23.

Williams, Bernard (2002), *Truth and Truthfulness: An Essay in Genealogy*, Princeton: Princeton University Press.

Williams, Michael (2001), *Problems of Knowledge: A Critical Introduction to Epistemology*, Oxford: Oxford University Press.

Wilson, Edward O. (1975), *Sociobiology: The New Synthesis*, Cambridge: Harvard University Press.

Wilson, Edward O. (ed.) (1988), *Biodiversity*, Washington: National Academy Press.

Wilson, Edward O. (1992), *The Diversity of Life*, Cambridge: Harvard University Press.

Wilson, Edward O. (1998), *Consilience: The Unity of Knowledge*, New York: Knopf.

Wilson, Robert A. and Frank C. Keil (eds) (1999), *The MIT Encyclopedia of the Cognitive Sciences*, Cambridge: MIT Press.

Wimsatt, W. K., Jr [1946] (1954), *The Verbal Icon*, Lexington: University of Kentucky Press.

Wolfe, Alan (1993), *The Human Difference: Animals, Computers, and the Necessity of Social Science*, Berkeley: University of California Press.

Wolfe, Cary (2003), *Animal Rites: American Culture, the Discourse of Species, and Posthumanist Theory*, Chicago: University of Chicago Press.

Wolfe, Cary (ed.) (2003), *Zoontologies: The Question of the Animal*, Minneapolis: University of Minnesota Press.

Wright, Robert (1994), *The Moral Animal: The New Science of Evolutionary Psychology*, New York: Pantheon.

Wylie, Alison (1992), 'The Interplay of Evidential Constraints and Political Interests: Recent Archaeological Research on Gender', *American Antiquity* 57, pp. 15–35.

Young, Allen (1995), *The Harmony of Illusions: Inventing Post-Traumatic Stress Disorder*, Princeton: Princeton University Press.

Zammito, John H. (2004), *A Nice Derangement of Epistemes: Post-Positivism in the Study of Science from Quine to Latour*, Chicago: University of Chicago Press.

Index